Metals Technic

A Collection of Techniques for Metalsmiths

Edited by Tim McCreight

tech·nic (tek'nik) *n.* 1. *Plural.* The theory, principles, or study of an art or process. 2. *Plural.* Technical details, rules, methods, or the like.

The American Heritage Dictionary,
New College Edition, © 1980.

Acknowledgments

This book is the direct result of the scholarship, enthusiasm and generosity of the authors. They have been consistently supportive and resourceful, and it is to them that primary credit must go. For the important work of proofreading, thanks go to Lynn Brunelle and Cynthia Selinger, whose editing care have clarified the language. This book was designed by Richard Mehl, who brought clarity and grace to both the material and the process of assembling it into a volume.

And on behalf of all of the authors, I want to thank the many students and friends who have worked with us over the years. The information in this book is largely based on workshop presentations, and in this sense it has come from our students. It is most appropriate, then, that we give it back.

Brynmorgen Press
33 Woodland Road
Cape Elizabeth, Maine, USA

Drawings by Tim McCreight

Book design by Richard Mehl,
Melilli.Mehl Graphic Designers

Photographs by Mary Melilli

The photographs that open each chapter document objects created by the authors using the techniques they describe.

Printed in the Hong Kong, China.

ISBN: 0-9615984-3-3
Library of Congress Catalog Card Number: 92-73596
Third Printing

Contents

INTRODUCTION

The following twelve articles represent the research, creativity and generousity of a dozen metalsmiths. In most cases this information was gleaned from handouts used in workshops, rewritten and illustrated specifically for this book. The information presented here is a composite of years of research and teaching experience, presented with the hope of making your own experimentation more rewarding.

Most of the information in this book is technical. It covers the *how* and explains the *why*, with-out asking questions about *what* is being made. This is not because the authors consider technique the most valuable aspect of metal-work, but because we feel that proper technique is a prerequisite to work of quality. We know that metalsmithing has always been intimately bound up with the materials and processes with which we work. The hammer truly is connected to the heart, so skill with one will benefit both. The following articles are small lights on a great path, not intended as beacons to mark an ending point, but as simple markers to light the way for whatever comes next.

John Cogswell

Sterling Granulation

The term granulation refers to a fusion process in which many tiny spheres are joined to a precious metal surface with a fillet so minute that it is almost invisible to the naked eye. The effect is of a pattern of grains that seem to be simply laid into place.

Granulation was known to many ancient cultures and was perhaps brought to its zenith by the Etruscan goldsmiths between 500 and 600 BC. Virtually all granulation that has survived from antiquity is worked in gold, with high karat yellow alloys being preferred. In the traditional method for that alloy, a trace amount of a metallic compound is introduced at the point of contact between the granule and the parent metal. Heating and subsequent chemical reactions create a localized alloy of a lower melting point, which creates the fusion weld.

In my own experiments I have developed a similar method for use on sterling silver. The fusion of this alloy does not require the addition of another metal, but rather depends entirely on the bonding properties of the sterling itself. By creating a fine silver skin on the granules, a range of melting point difference is established that will allow the fusion to take place with reasonable control. Granulation is more a matter of experience than knowledge. No amount of reading will substitute for the observations and judgement that you will gain by experimenting with the process.

Tools, Supplies and Equipment

Most of the tools and supplies needed for this process are standard items in a metalworking studio. A rolling mill and a small kiln are useful and time-saving, but are not essential.

- Presto-lite torch with #1 (or 1A) and #3 tips
- charcoal block
- liquid hard soldering flux
- scissors
- fine pointed tweezers (eg. watchmaker's)
- cross-lock tweezers
- sable brushes (000 and 00)
- pickle (sulfuric acid solution or Sparex)
- brass scratchbrush
- ovenproof glass dishes and pie plate
- optional: kiln, graphite crucible, gloves and tongs
- shallow dish or large spoon
- sieves and containers for storing granules

Making the Granules

Make granules from sterling silver foil; about 30 to 36 gauge. If a rolling mill is available, the foil can be made by rolling out small scraps of clean (no solder) sterling sheet, gradually tightening the rollers until they are completely closed. Several more passes through the mill at this point should yield a foil of appropriate thickness. If the scraps of sheet are annealed before the rolling begins, no subsequent annealing should be necessary. Cracking at the edges poses no problem because the foil will later be cut into small pieces. If no rolling mill is available, purchase sterling sheet as thin as you can, or forge sheets on a clean anvil. Slightly heavier gauges work equally well; they are simply harder to cut up.

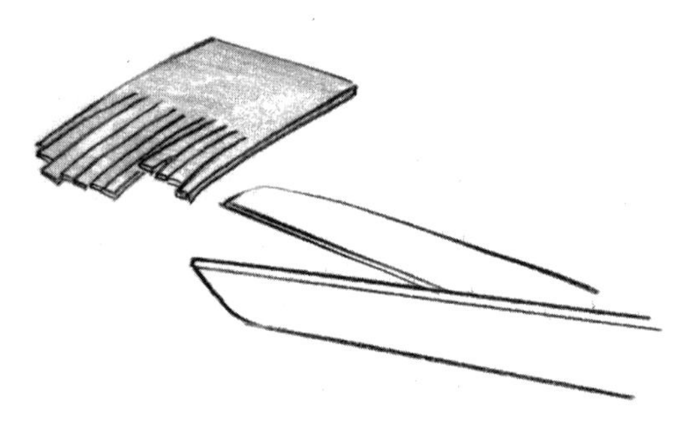

Roll clean sterling sheet until very thin and snip into small pieces by making a fringe and cutting across it.

Clean the foil well, then cut it into tiny snippets with scissors exactly as you would cut sheet solder. Snippets should be about 1/16 inch square if 36 gauge is used. Because the pieces will not be perfectly uniform, they will yield a range of various sized granules.

Make some provision to capture the snippets as they fall. When making a large supply of granules, clamp one handle of a scissors in a vise as shown here. A large sheet of paper placed directly beneath the vise will catch the snippets of foil. When enough chips have been cut (you'll want a large supply; be generous), they are ready to be melted into spherical grains. This can be accomplished either by using a kiln or with a torch on a specially prepared charcoal block.

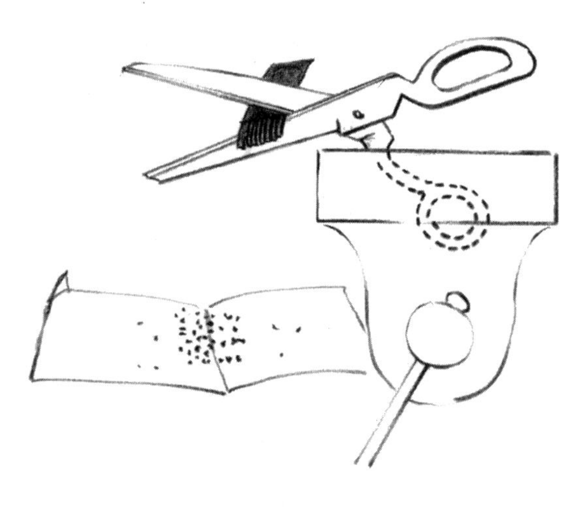

Spread a piece of paper beneath a vise to catch the snippets.

The Kiln Method

The kiln is preheated to 1900°F (1038°C) and adjusted to maintain that temperature. In the meantime, prepare a quantity of charcoal dust or powder sufficient to fill the graphite crucible by filing or pulverizing old charcoal blocks. Use a coarse file over a drop cloth, or set the chunks of charcoal in a heavy paper bag and pound with a mallet. Rather than waste a good block for this, use pieces of old blocks or hardwood charcoal barbecue briquets. This is best done out of doors, as it can be a messy procedure.

Place a ¼ inch layer of this powder in the bottom of the crucible, and over this sprinkle a small number of the silver snippets. The snippets should be distributed as evenly as possible, because chips that touch will eventually melt together. Alternate layers of charcoal and snippets until the crucible is filled to within a quarter of an inch of the rim, the top layer being powdered charcoal. Lay several pieces of stout iron wire (e.g. iron coat hanger wire) across the mouth of the crucible. On top of this, place a lid of unglazed ceramic slab or a slice of firebrick.

Lay alternating layers of powdered charcoal and silver snippets into a crucible.

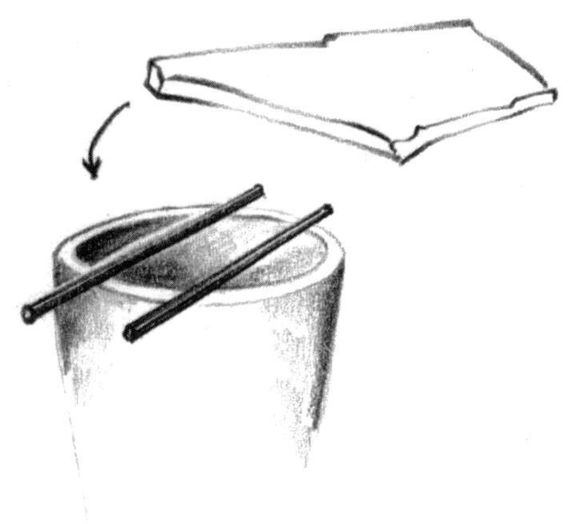

Make a lid for the crucible by setting a piece of ceramic or firebrick across pieces of coathanger or welding rod.

When the kiln has reached 1900°F (1038°C), carefully insert the covered crucible with the crucible tongs. Heat-proof gloves should be worn, of course, to avoid burns. The filled crucible must be handled very gently to avoid having all the snippets settle to the bottom. If this occurs, the end result will be a solid slug of silver instead of numerous tiny granules.

After the kiln has come back up to temperature, the crucible is allowed to soak for about 15 minutes. The actual time necessary to completely melt the chips will depend on the amount of metal in the crucible and the size and thickness of the crucible itself. After 15 minutes, remove the crucible from the kiln and scoop out a sample from the top layers with a spoon. If the top layer of snippets has completely melted into spherical granules, chances are that the lower layers have melted also. If melting is incomplete, the crucible must be recovered, replaced in the kiln, and allowed to soak longer. Check the progress of the contents every 10 minutes or so, until all the chips have melted into granules.

When the melting is complete, remove the crucible from the kiln and dump the charcoal and granules from a height of 10-12" into a Pyrex bowl containing at least 3" of water. This may be done when the contents have cooled to a dull red heat, but do not quench the crucible itself. Run a gentle stream of tap water into the bowl, floating the charcoal powder and ash residue off with the overflow. When all this residue has been flushed off, the water in the bowl is carefully decanted off, leaving behind an assortment of perfectly spherical granules. Transfer these granules to another Pyrex bowl containing the pickle. The granules should remain in this solution until they acquire a uniformly matte white surface. Pickling can be hastened if the pickle solution is warmed, but do not heat to boiling.

When the granules have been thoroughly pickled and the solution has cooled, carefully pour it into a storage container and flush the granules in water to wash off any residual acid. Drain as much excess water as possible from the bowl. Place the granules in a pouring vessel or spoon and gently heat from below with a soft flame. Use only enough heat to evaporate the remaining moisture. When the granules are dry and roll around freely in the bottom of the spoon, remove the flame and the allow grains to cool. At this point they are ready for use. If the granules are to be sieved, it is done at this time and the various sizes stored in separate containers.

Dry the washed grains in a spoon over a mild heat.

When making spheres with a torch, tie a copper sleeve around a charcoal block to prevent the grains from rolling off sideways.

THE FREE FALL METHOD

The second method by which the granules can be made requires only a charcoal block and a torch for the melting sequence. Though requiring less in the way of equipment, it is far more time-consuming, because only a small number of granules can be melted at a time. When I use this method, I like to fit the charcoal block with a special metal collar to help contain the rolling granules.

Cut tiny chips of clean sterling as described earlier. Sprinkle these on the charcoal block that has been propped at a slight incline, the gate facing down, about 12" above a Pyrex bowl full of water. Using a small tip (Presto-lite #1 or #2), melt each snippet individually, starting at the lower end of the block and working up. As each chip draws up into a sphere, it rolls off the block and into the water. The metal collar prevents the granules from rolling off the sides of the block and funnels them into the bowl of water below. The sprinkling and melting steps are repeated until you've made an adequate supply of granules. These granules are then pickled, rinsed and dried as described on the preceding page.

The tilt of the block will allow the spheres to roll off as they form. Be sure that there is sufficient drop so they can harden before they hit the dish of water.

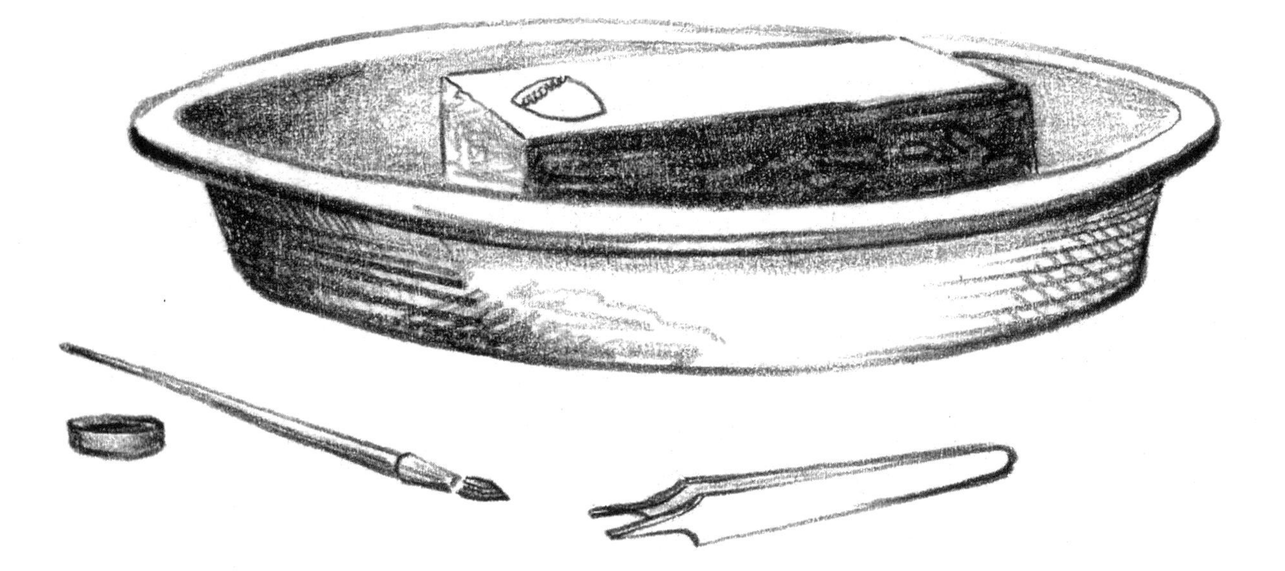

Setting the charcoal block into a cake pan or similar tray will make it easier to retrieve any fallen grains.

Preparing the Surface

The base, or parent surface to which the granules are to be fused should be between 20 and 24 gauge B&S in thickness. Thinner gauges overheat quickly and can collapse or melt before proper fusion takes place. Small discs, ½ to ¾ of an inch in diameter, stamped out with a disc cutter and slightly domed in a dapping block, make ideal practice samples.

The base surface must be properly prepared before the granules can be applied. Granulation performed on sterling silver always yields a slightly roughened surface. This torch texture is minimal (often hardly discernible) when the surface has been pre-finished to a high rouge polish; a coarse pumice, or emery paper finish results in a much more pronounced surface texture. This texture resembles that achieved by sand-blasting, and can be quite attractive. The surface should be free of nicks and scratches, because their removal after the granules are fused in position is virtually impossible.

When the surface has been pre-finished to the desired degree, the piece should be thoroughly scrubbed with soap and water to remove any polishing compounds, oils, etc., and then alkalized with ammonia or saliva to eliminate surface tension. When rinsed, water should sheet evenly over the surface. If water beads up, the cleaning and alkalizing must be repeated. Dry the piece with a clean tissue and handle only at the edges to avoid fingerprint contamination. Place the work on the charcoal block and set this in a Pyrex pie plate, as shown.

Applying the Granules

To pick up the granules for transfer to the surface, lightly moisten the #1 brush in the flux solution and dip it into the container of granules. Deposit the granules onto the surface with a rolling motion of the brush. I find it most efficient to transfer a quantity of granules to the surface before sliding them into position.

Working from one end of the design, position the granules with the #000 or #00 brush and fine-pointed tweezers. If the flux solution begins to dry too soon, add a drop of distilled water with the brush. Additional flux solution should not be used, because excessive flux build-up may cause boil-up and displacement of the granules during later heating. When the pattern is completed, any leftover granules are coaxed off the surface, and off the charcoal block into the pie plate from which they can later be retreived and reused. Remove excess moisture by blotting with the tip of a dry brush. Allow the piece to dry thoroughly, which typically takes 15 minutes.

FIRING

Set the workpiece near a corner of a charcoal block to allow yourself room on the block to adjust the torch flame. Using a #3 tip, adjust the flame at the needle valve until only a slight tinge of yellow shows at the end (the same type of flame used for soldering). Hold the torch tip about 2" above the block, and reduce the pressure at the regulator until the flame produces a soft, mushroom-like corona about 1½ to 2" in diameter on the surface. These adjustments produce a reducing atmosphere within the confines of the soft corona.

Move the flame toward the piece slowly, watching closely for any movement or traces of steam. Work at eye level to insure the best possible visual access. If no movement is observed, play the flame directly over the piece with a steady on-off wafting motion. Until the piece reaches bright red heat, the surface should be kept within the corona as much as possible to inhibit oxide formation. Once this stage is reached, move the flame completely off the piece with each pass to allow even dispersal of heat and to prevent overheating of the granules.

At red-orange heat, the surface will begin to sweat, or become shiny and mirrorlike. This is the point at which fusion occurs. Because the fillet formation appears as a mercury-like wink or flash only at the base of each granule, it is crucial to positon the work at a height that allows the fusion to be clearly observed.

Fusion occurs at localized areas so the piece must be rotated and the heating continued until fusion has been observed over the entire pattern. I recommend that the flame be moved completely off the surface, and the piece be allowed to drop back to bright red heat before you progress to a new section. Sustained heating at this point may cause collapse or partial melting. When the firing is completed, check the bond formation by gently tugging at a few granules with fine-pointed tweezers. If this test proves satisfactory, turn off the torch, set the piece aside to air cool and extinguish any glowing areas on the charcoal block with a damp rag. Do not quench the finished piece because steam thus generated may dislodge the granules.

At this point, granules and base have become a single, homogenous unit. After pickling, the piece can be safely soldered without fear of remelting the fillets. If fact, it has been my observation that repeated heating actually seems to improve the quality of the bond.

Cleaning and Finishing

When cooled, clean the piece in a standard sulfuric acid or Sparex pickle, rinse it in a baking soda solution to neutralize acid that may be trapped in the tiny spaces between the granules, and rinse it again in clear water. Gently burnish the work with a brass scratch-brush, using liquid detergent and water as a lubricant. The brush will impart a fairly high luster to the granules. If a higher polish is desired, hand polish the granules with a rouge cloth after brass brushing. Under no circumstances should the granulated surface be subjected to the action of a motor driven wheel. Even with rouge, the cutting action is so fast that the tops of the granules quickly become flattened, destroying their spherical integrity. In addition, the granules may become snagged by the fibers in the buff and pulled off the surface.

The granulated pattern can be oxidized in a liver of sulfur solution, just like any other sterling silver item. The color can be rubbed from the tops of the granules with a thumb dipped in fine pumice, and the piece then brass brushed; the visual aspect of bright granules against a darkened background can be a very effective device for enhancing the granulated pattern.

Some Common Problems

Carburization of Granules

When granules are made by the kiln-crucible method, they often develop a blackened surface that is not readily removed by the pickle. This blackened surface is the result of carburization, or absorption of carbon, due to prolonged heating in the carbon rich charcoal powder in the crucible. This usually occurs when the crucible is placed into a cold kiln and then brought up to heat. The extended, gradual heating allows carbon to unite with the silver, and results in a surface deposit of metallic carbide that is highly resistant to the action of the pickle. To avoid carburization, place the crucible in the chamber only after the kiln has been preheated to the proper temperature, and then leave it in only long enough to melt the snippets into granules.

If carburization has already occurred, the granules must be repeatedly heated and pickled in a warm sulfuric acid solution until they are a dead white color. Simply leaving the granules in the acid pickle for a long time is not sufficient to remove the surface carbide layer, and prolonged immersion in this strong solution exposes the silver to the corrosive effects of the acid. At least two or three heating and pickling cycles are needed to remove the carburized surface. The granules need to be heated only to about 400-500°F (204-260°C) for this purpose. A small piece of paper placed in the vessel beneath the granules provides a convenient temperature indicator. When the paper begins to turn brown (450°F, 233°C) the granules will be sufficiently heated and should be immediately poured into the warm pickle solution. Leave them in the acid solution no longer than 2 or 3 minutes between each heating.

Oxidation

Oxide formation on the base surface during the initial stages of heating is usually an indication of improper flame adjustment, or failure to keep the surface enveloped within the reduction atmosphere of the corona until the bright red heat is reached. Some slight oxidation is to be expected; the flux at the base of each granule is usually sufficient to handle this. When oxidation is excessive, however, the only recourse is to repickle the base and granules, and start all over again. Because excessive oxidation inhibits or totally prevents proper fillet formation, continued heating is a waste of time and potentially disastrous. The temptation to induce fusion by the application of "just a little more heat" usually results in collapse or melting.

Overheating

There is no remedy for meltdown; prevention through proper heat control is the only solution. Continued heating beyond the sweating stage will result in rapid successive stages of breakdown, the final step being complete melt. This is typically followed by gnashing of teeth and bad language.

The first stage of overheating is indicated by deformation of the base piece. This is most noticeable on domed or otherwise formed pieces, as overheated areas slump or sag. On flat pieces that are supported by the charcoal block, the effect is less obvious, appearing only as slight warpage.

The next stage of overheating affects the granules themselves. The fine silver skin is no longer sufficient to hold the grains in spherical form and they begin to diffuse, eventually becoming amorphous mounds on the surface. At about this same point, the edges of the base piece begin to pull in and finally the piece melts completely. The time span from sweating temperature to complete melt is only a matter of seconds.

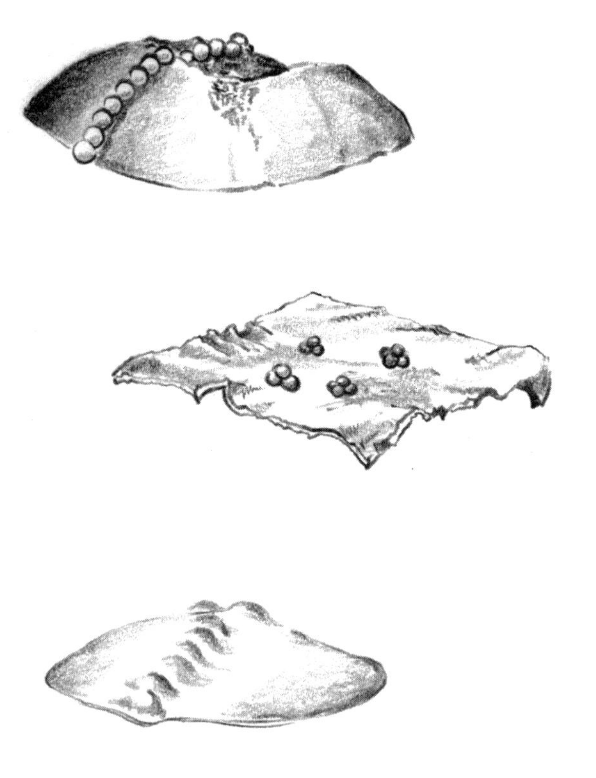

Slumping, warping and absorption are three problems with the same cause – overheating.

Torch Texture or Blistering
Some surface texture is expected when granulating on sterling. However, under certain circumstances, the surface can become severely textured and/or blistered. In most cases, this undesirable deformation of the surface is due to one or more of the following factors:

- general or localized overheating of the surface
- excess flux
- fingerprint or other contamination of the surface
- excessive oxide deposits which may develop during initial heating
- allowing the intensely hot inner blue cone of the flame to come too close to the surface

Incomplete Bonding
Even under the best circumstances, one or more granules in the pattern may not bond when all others have. If this happens, simply pickle and clean the piece, re-apply granules as necessary, and re-fire. If none of the granules have fused, excess oxidation or insufficient heat are usually to blame. The only recourse is to start over again.

Weak or Porous Bonds
Weak, porous or partial fillet formation usually results from surface oxidation or insufficient heat application. It's important that the sweating stage be maintained long enough for the "wink" of bonding to be clearly observed.

Problems Setting Up the Pattern
One of the most tedious (and exasperating) aspects of granulation is the set-up, or pattern development. At times, the granules will behave as though they have a mind of their own. Capillary action causes the granules to pull together into characteristic formations, three-granule triangles and four-granule pyramids being most common. If the granules are too dry they are hard to move. If they are too wet, the addition of one final granule to complete a pattern may shift the entire mass out of position. Careless handling of the piece after drying can cause the pattern to suddenly disappear from the surface altogether. Patience and a deft touch will overcome most of these problems. Take occasional breaks to relieve the tension of concentration and use an Optivisor® to reduce eyestrain.

John Cogswell is a metalsmith and jeweler who combines teaching with an active production schedule. He is a program director at the 92nd Street Y in New York City and teaches dozens of successful workshops around the country each year.

Phillip Fike

Niello

Niello is a lustrous substance that can be made by the metalsmith and is generally applied as a decorative inlay, in some ways akin to fused glass enameling. While technically it is not a metal, niello looks, melts, cuts and polishes like a metal. It holds a prominent place in the history of art as a unique material, so special that nothing else can take its place.

Niello is made by melting copper, silver and lead together in a crucible, then adding sulfur to the alloy to create a mixture of sulfides. This brittle material has a color that ranges from blue-grey to black, and has been used since antiquity to enhance decorative objects made of gold and silver. It is neither malleable nor ductile, but becomes fluid at a relatively low temperature and adheres well to precious metals. Once fused to a parent material, niello can be polished to a brilliant finish by standard metalworking processes.

The origin of niello is unknown, possibly prehistoric, and probably discovered by accidental alchemy. Early on it is seen throughout Europe, Egypt and the Middle East, and most prominently in the Far East, specifically in Thailand. It became a popular inlay substance because of its ease of use, the availability of the ingredients and its ability to enhance contrasts between light and dark values inherent in the patterns of decorative objects.

In Renaissance Italy, anything pictorial in all the arts was active and thriving. During this period, goldsmiths were engaged in engraving metal plates with the designs and drawings of other artists who lacked the technical skill to engrave the plate. It's interesting to realize that before the 15th century, the knowledge and practice of lifting a paper print from an engraved metal image was not yet understood. Even though chemical etching and mechanical engraving were widely used to cut lines in metal, the possibilities for printing were yet to be realized.

It is natural for the engraver to check the progress of his work by rubbing a dark paste into the gravure. What was missed for so long a time was the possibility of rubbing the surface clean after the plate was fully inked, leaving the grooves filled. When a wetted paper was laid upon a plate prepared in this way and subjected to sufficient pressure to push the paper into the grooves, an exact replica of the plate was produced in ink on the paper. Exact with one important distinction: the image is reversed.

My research has convinced me that engraving as an image making process of goldsmithing broke the ground leading eventually to etching and intaglio printmaking. Both methods have flourished since the 16th century and nielloworK holds a prominent position in their development.

It is only natural to assume that artists of the period quickly understood the potential marketability of prints as multiple items. "Why settle for a single object when it is possible to have many!" This thinking lead to a distinction being made between plates made for the purpose of printing and plates intended as final objects in their own right.

The relationship between the goldsmith-engraver-printmaker was a short-lived phase, lasting about a hundred years. The popular iconography of the time included numbers signifying a particular date, slogans, mottos or maxims, which in the plate would read correctly, but would be reversed in a print. It soon became generally understood that all engraving with the iconography reading correctly in the print meant that the plate was scheduled for multiple proofing. The engraver-goldsmith whose plates were intended to be one-of-a-kind metal plates filled the gravure with niello, permanently fixing the image and making it unprintable. Such an artist came to be called a "niellist," and his works were called "nielli." By the 18th century, these "nielli" were scarce indeed, and had become highly sought by connoisseurs and collectors, providing a climate and a market for forgeries.

Equipment and Materials

The Fire System

In 500 BC I would have fashioned a fire pot with bellows attached to a clay pipe, strong enough to force air directly into the burning fuel. If you have a passion for ancient methods, something like this is as possible today as it was then. Consider digging a hole in firm soil, about 15" in diameter and of the same depth. Burrow a 2" diameter hole from the side, angling downward and entering the lower region of the fuel chamber. Wood, charcoal, charcoal briquets, green pea coal, coke or "prairie pancakes" can all be used as fuel. If the air feed hole collapses, reinforce it with a length of steel pipe. Attach this to a two-chambered bellows (which you'll probably have to make) or a handcrank blower, vacuum cleaner motor or a hairdryer.

In my properly equipped studio, I make niello using an oxy-acetylene torch, which provides all the heat required to bring copper to a quick melt. In a workshop situation, I have students work with a "Presto-Lite" acetylene/air torch fitted with a #5 tip. Alongside I assist with the oxy-acetylene unit to move things forward rapidly.

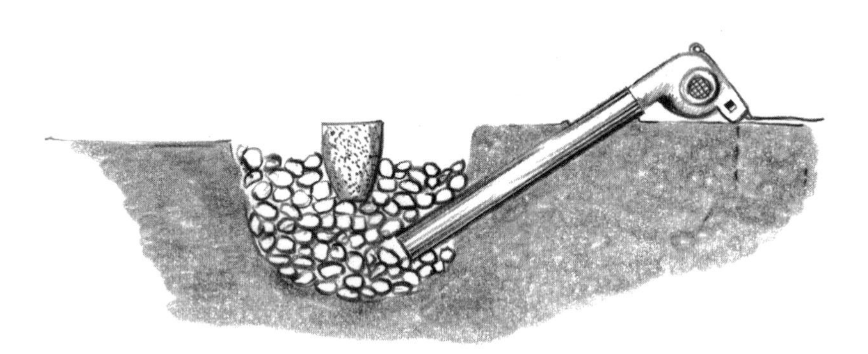

An open pit hearth, with a few modern improvements.

Fume Hood and Exhaust System

Dangerous fumes from metals, fluxes and burning sulfur can be controlled in a temporary fume hood fitted with a common household fan. A laboratory fume hood is ideal, and a coke forge with a strong updraft will work well, but if you have neither of these on hand, here is a jerryrigged model that has served me well in the past.

This drawing illustrates an effective hood I have made from cardboard and cheap lumber. Start by locating a window that will provide an appropriate exit for the fumes, preferably away from people or a high traffic area. Set a 24" fan into the window, often simply sitting it on the windowsill. If necessary, rig up a shelf or brackets to guarantee that it is secure.

When the fan is in position you can decide the dimensions of the hood itself, which might in turn be determined by the size of the window frame. Keep in mind that a smaller hood means less air to move and therefore increased efficiency. The front may be left open all the way across, or may require a larger panel with an opening as shown here.

I've made many of these hoods as I conduct workshops around the country, and have had good luck with a simple hood made from standard corrugated cardboard. While a simple box can be used, a sturdier hood can be made to an exact size by first creating a frame from wooden strips (lath or furring). This is expediently cut and nailed together, then a skin of cardboard is tacked and/or glued over it and sealed with duct tape. Materials needed will include about a refrigerator carton's worth of cardboard, a hammer and 1" carpet tacks, shears, a matt knife, a glue gun with several sticks of glue and a tape measure. The whole affair can usually be assembled within an hour. Though I doubt you could convince a fire marshall of the wisdom of creating a fume hood of combustible materials, in literally hundreds of hours of use, I've never seen one set afire. I enlist your common sense and safety consciousness to perpetuate that record!

Check the strength of the draw by setting a tightly wadded ball of newspaper into the hood. Light it then blow it out to create a smoky environment. If the smoke is not immediately and entirely drawn outside, make the hood smaller and check the tightness of the joints. This is going to be a chamber filled with hellish fumes and you can prevent them from reaching your lungs by simple ingenuity. I have always regretted having to tear a good hood down.

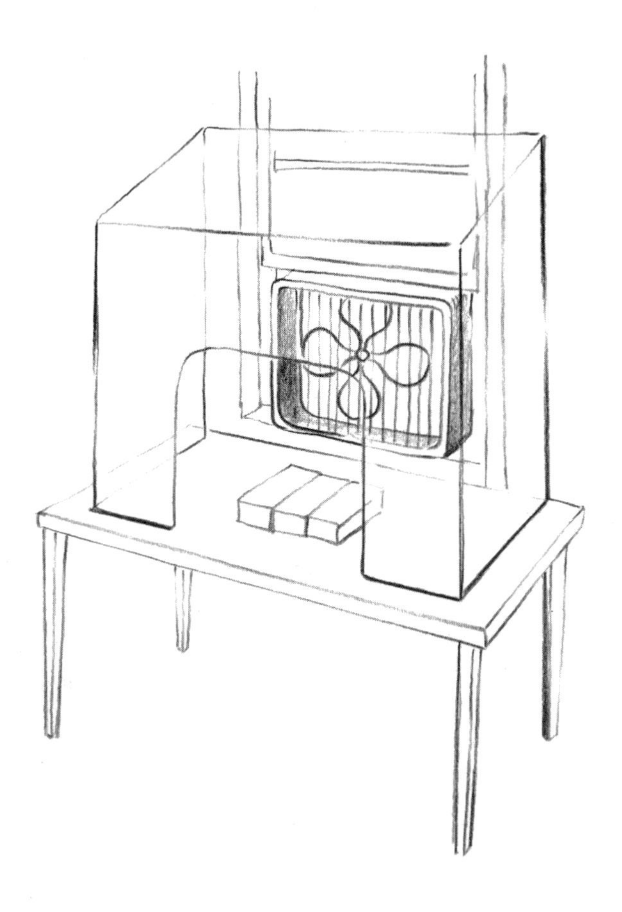

This transparent view shows the installation of a simple yet effective fume hood. It is made of cardboard fixed to a wooden frame and powered with a standard household fan.

The Work Table
Work centers on a sturdy table with a fireproof surface. A layer of bricks on a wooden table will be sufficient. A standard kitchen counter height will be most comfortable. The crucible will rest on 3 bricks set into the center of the hood; what I call *Crucible Island*. This placement has the advantage of keeping the work up where it is easier to reach and allows you to move the working center further into the hood if the draw requires it.

Crucibles
Crucibles made of fused silica clay, high fired ceramic refractory and graphite will all work. My favorite is the largest "Burno" brand, which has a 100 pennyweight capacity. Its removable cover helps collect heat, providing important control just when it's needed.

Iron Tongs
The grasping ends should be forged and shaped like fingers so you can achieve a sure and comfortable grip on the crucible. Niello-making requires sure handling and control, so the tongs must allow for the same kind of dexterity you have with your bare fingers. There simply isn't time to be fumbling around, nor it is safe, because of the possibility of spilling the crucible. One hand will be busy holding the torch, so the other must be adept. Shape the tongs and practice with them until you can pick up the lid and the crucible without a moment's hesitation.

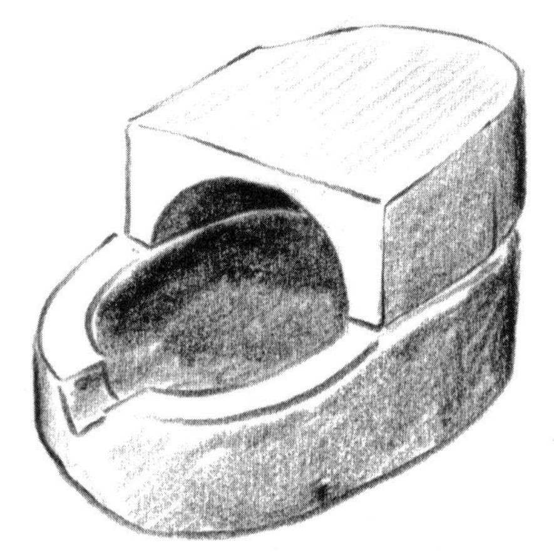

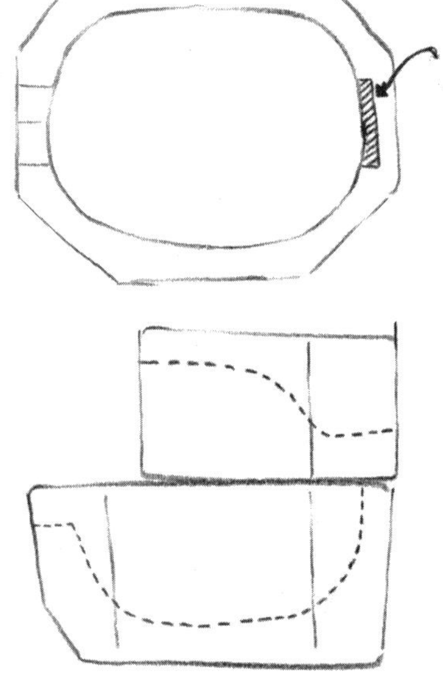

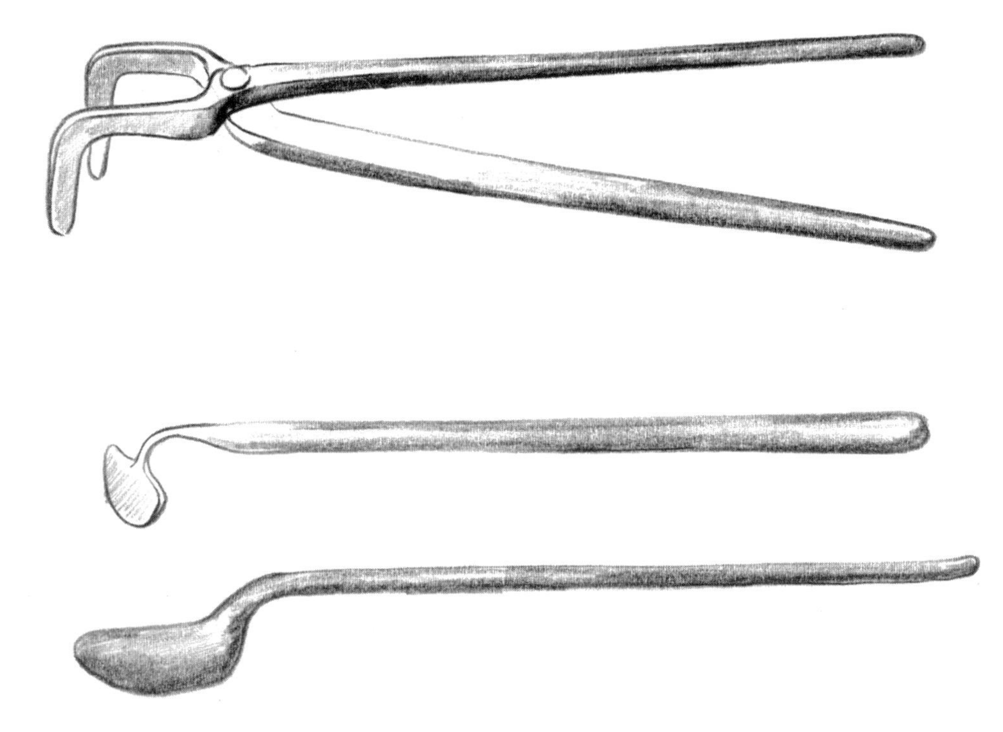

Tongs like these can be bought or made,
and should be adjusted to allow
a comfortable grip of the crucible.
The scraper and muddling tool are forged
from steel rod to suit your needs.

LEFT

This is the style of crucible I prefer,
though others can be used.
Before the first use, I grind a flat area
at the back of the lower unit to
facilitate a positive grip with the tongs.

Iron Stirring Rod and Muddler

In the process of making a batch of niello, the crucible will need to be scraped and the mix stirred. This can be achieved with a single plain rod, but the two functions are better handled with rods designed for each purpose. Start with a pair of rods about 2 feet long and at least ¼ inch in diameter. The shapes illustrated are examples of typical forms but you can devise your own to meet your needs. Remember that these rods should be reserved for this use: you don't want traces of niello showing up in your casting or soldering operations where it will contaminate sterling or gold.

Ingot Mold

In my work I use niello primarily from a "Splendid Rod," though some niellists prefer to work from a powder in a manner similar to enamel. To make this, niello is poured into water, into a standard ingot mold, or onto a surface plate, whence it is broken into pieces and ground to a powder in a mortar and pestle. I prefer to pour the molten niello into a length of "smoked" angle iron, positioned at a slight incline as shown. After using several types of molds I have settled on standard 1 or 1½" steel angle iron mounted on blocks. For standard studio work I use a mold about 4 or 5 feet long, though I have used a 12 foot length and have poured a Splendid Rod 10 feet 2 inches long.

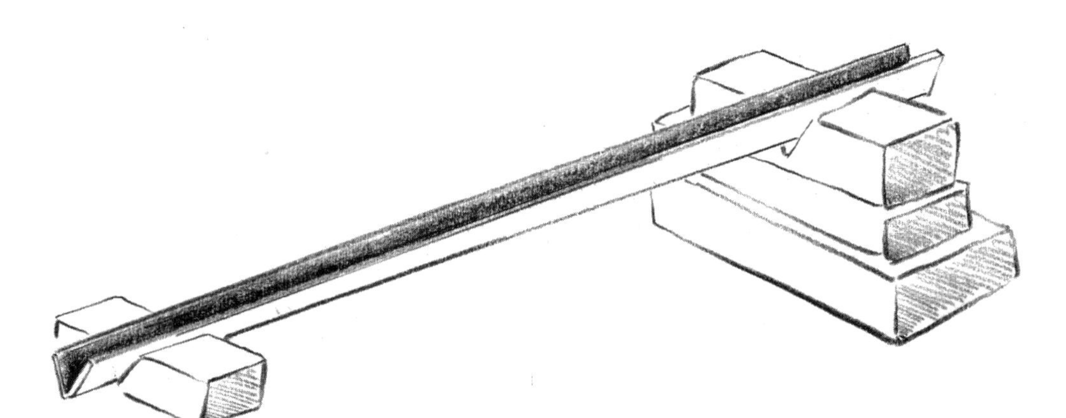

An angle iron is set on bricks or blocks of wood to create a slight decline. This becomes the ingot mold for the "Splendid Rod."

Miscellaneous Tools

- a scale or balance
- bottles for sulfur and flux, and long handled spoons for each. These are easily made by brazing a rod to a stainless teaspoon.
- a small box of sand to securely hold the bottles at a convenient angle
- small open mouth bottles for the metal components
- tweezers to lift the rod from the mold

It is good practice to layout all the tools and materials before starting the process. In fact I recommend a dry run to insure that the tools are comfortable and that the position of all the ingredients and tools is correct for efficient operation. The ingot mold should be close at hand, so you can move there in a single step while keeping the torch on the crucible.

RECIPES

Written material about niello is scarce and sometimes contradictory. As a student I was fortunate to have access to H. Wilson's *Silverwork and Jewelry*, whose chapter on niello contains several recipes. Wilson apparently tried many of the variables and ultimately selected an alloy of 6 parts silver, 2 parts copper and 1 part lead. After my own experiments I have settled on a recipe that includes an extra measure of lead, namely:

6 parts fine silver
2 parts copper
2 parts lead

Two heaping tablespoons of sulfur complete the recipe, being sufficient to convert these metals to their sulfides. Because many things can go wrong until the process is under control, I originally recommended that the beginner start with small quantities. I have changed, believing that larger batches are both easier and more useful. I personally start with 12 grams of silver (therefore 4 grams each of lead and copper).

I prefer to use fine silver, oxygen-free copper and pure lead, but I have made niello from sterling silver, scrap copper, lead from a plumbing supply and sulfur from a garden nursery. If your process is good these materials will produce a successful batch.

In like manner, it isn't necessary to be exact in weighing the ingredients. A half a gram more or less doesn't seem to matter. The key factor for success seems to be in how you use the fire; it is critical to keep the alloy hot enough without overheating it.

If you are using a crucible for the first time, it should be coated with a layer of flux. Heat the crucible to a clear red, then sprinkle the interior with a borax flux. Repeat the process if necessary to insure that the crucible has a uniform coating of a clear glassy flux layer.

Introduce the copper and bring it to the molten state. Drop in the silver and continue to heat the crucible. Stir the alloy to insure a thorough mix of the two metals. Make a mental note that from this point forward it is possible to overheat the alloy or to take too long in completing the process.

Introduce the lead and again pick up the crucible and swirl it to encourage the elements to mingle. Two rotations is sufficient. This phase requires sensitive heat control and virtually no hesitation. Place the crucible on its island and turn on the exhaust system. With one smooth and swift delivery, introduce a heaping tablespoon of sulfur into the alloy. The sulfur is instantly afire, creating a dramatic display and a lot of smoke, which the exhaust hood will take away. A timid delivery may accidently set afire the spoon as well, which you might return to the bottle unknowingly. Not good.

Pick up the stirrer (muddling rod) and push, drag, stir and generally 'muddle' the burning sulfurous liquid into the alloy. The burning sulfur can be splashed if the action is too vigorous and a nasty burn could result; be gentle until you get the hang of it. Protective goggles, gloves and an apron are recommended.

At this point the niello will be about 1200°F (650°C). For a brief period you may see it as a medium red mass swimming in the sulfur that is burning around it. Add an additional (smaller) amount of sulfur and continue the mixing. The alloy will quickly become saturated with sulfur and any more added at this point will simply burn away, prolonging the process to no advantage.

Pick up the crucible with the tongs and give both the niello and the sulfur residue intermittent blasts with the torch. This keeps the niello molten and burns off the excess sulfur. At this point it is very important to swirl the niello in a circular motion around the walls of the crucible. This allows a residue to separate itself and gather in the center of the crucible as a cinder, awash in the rotating niello liquid. If you overheat the alloy at this point the cinder will melt and rejoin the liquid mass, but this would be a big mistake. The cinder seems to be a purifying agent, attracting to itself some of the ingredients that don't join in the sulfide solution. The cinder might account for a fourth or a third of the total mass of the niello.

Use the torch in a back and forth motion to sustain the crucible temperature. In demonstrations assisting a student, I use the word "zap" to indicate a split second thrust of the torch directly into the crucible. The idea is to keep the niello molten but to avoid the damage of overheating. Eventually a rhythm of control between fire and metal becomes second nature.

If the alloy gets too hot, the cinder will melt back into the alloy. If it gets too cold the niello will not be liquid enough to pour. This is not a terribly difficult temperature to maintain, but there is no doubt that you will get better with practice. Move to the ingot mold and line up the crucible with the axis of the angle iron. With the torch keeping the niello just liquid enough to pour, tilt the crucible forward and allow the alloy to flow into the mold. You must not be too slow at this, but neither should you splash the niello into the mold.

In most cases, you'll create a rod between 10 and 20" long. Return the crucible to its island on the worktable and replace the cover. Reheat the chamber to a full red heat, melting the residue to a viscous liquid. Tilt the crucible to a vertical position and pour out the residue, assisting the cleanup with the iron scraper. The crucible is now clean and heated throughout, ready to proceed directly to a second batch. The niello/borax residue that was removed can be saved for future remixing and experiments if you are so inclined.

After you've made 3 or 4 batches, break up the individual rods into pieces about an inch long. For the second phase of the process, melt all these pieces together. If you have a new crucible on hand, coat it with flux and use this for optimum results. The cleaner environment at the walls of the crucible will have a noticeable effect on the behavior of the niello. Repeat the melting as before but do not add any metal, flux or sulfur. If a small amount of residue appears, do not return the cinder to solution by overheating. Swirl as before, and return again to the ingot mold. Create another pour, this time resulting in a "twice poured" rod of exceptional purity: the Splendid Rod!

APPLICATIONS FOR NIELLO

Niello may be applied directly on a surface of gold or silver as a dark field or background. Niello is opaque at 2 thousandths of an inch. At .0005" it becomes slightly translucent with a brownish cast. Perhaps the most common application of niello is as a fill for lines and patterns made by engraving, chiseling, stamping, carving, chasing, etching, machining, electroforming, rollprinting, fabrication (channelwork) and casting.

All soldering and brazing must be completed on the workpiece before the niello is applied. The work is carefully cleaned by abrasive scrubbing, pickling, sandblasting or electrocleaning. The area to receive the niello is fluxed either with white paste brazing flux, diluted with water to 25-50% strength, or with Batterns flux at full strength.

Powder Method

Break pieces of the alloy with a mortar and pestle into chunks, gravel or grains, experimenting to determine your favorite size. Mix the niello powder with flux to make a thick paste and apply it to the work with a spatula or small brush. Flat or nearly flat objects can be fired in a kiln. As a liquid, niello is very quick to move and will flow according to the laws of gravity.

Rod Method

Hold a rod of niello in your hand or with tongs or tweezers. I heat the fluxed piece to the temperature at which the flux makes a clear glassy coating (1100°F, 600°C). This is well over the melting point of niello. With the torch pulled away but keeping the piece hot, I then touch the niello rod to the workpiece, where the melting niello passes through the flux and adheres to the clean metal. A generous overfill is almost always necessary.

When melting niello on silver and gold, avoid severe overheating or prolonged heating, because at this point the sulfide-silver bond deteriorates, causing the sulfur to exit and the silver to recrystallize unpredictably in the niello. This can be a desired event because of the attractive silver dendrites (snowflakes) that are formed, but most of the time it's unwanted and troublesome, causing large clumps of silver to emerge where they interfere with the design of the work. This needs special research by someone lucky enough to discover the method for reliable and repeatable control of this phenomenon. Overheating is also responsible for a blanched color visible on the surface, leaving a slight film of grey silver where one expects the dark rich tonality of niello.

As the niello displaces the molten flux, some of it floats on the surface as islands of glass. With practice these blobs can be picked up, either with the niello rod itself or with a steel probe. If this proves too difficult, they can be dealt with later by mechanical removal when the piece is cold. These flux blobs can interfere with the leveling of the niello because their undersides have a miniscus curve that forms a crater. After the recess has been filled and trimmed, it will be necessary to flash-melt the niello to bring all surfaces to a common level. The flux may also be removed by standard pickling. Follow this with a very thorough rinsing to flush pickle from any pits.

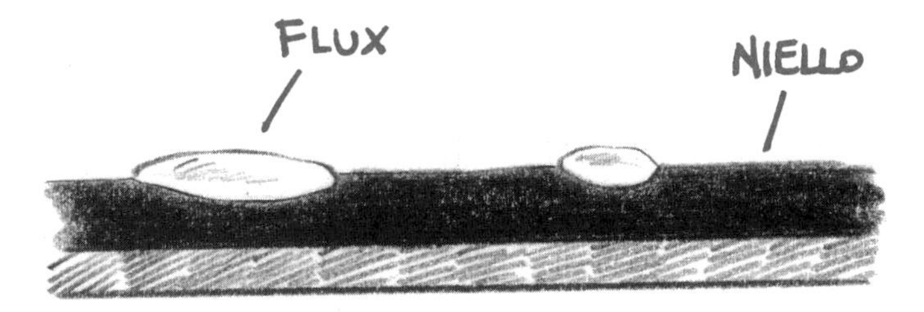

Small bubbles of flux can float on the top of the niello, creating craters with the underside of their miniscus curves. Overfill the section or pick off the flux blobs when the niello is molten.

Finishing Niellowork

Excess niello may be removed rapidly by milling, rough sanding and grinding, by heavy hand filing or by power cutting with burs in a flexible shaft. These are all rather violent methods of removal and I usually prefer a slower and more gentle removal style. In flat or slightly curved pieces, the best tool is a triangular scraper, with its edges honed as sharp as a knife. This will remove thin shavings that are clean and free of the steel particles that come with the use of files.

Remember that niello scrapings are valuable and can be reused if saved carefully. Though I always make fresh niello for special pieces, I collect all my scrapings for later experiments and my ongoing attempt to create an even longer "Splendid Rod."

As excess niello is removed and when the pattern below begins to appear, I immediately shift to another area not yet revealed, never going too deep into the pattern at a single spot. This process is continued throughout until all sections of the pattern are revealed. If I am working on a delicate engraving, I will at this stage abandon the scraper and take up a series of "wet or dry" silicon carbide abrasive papers, delicately abrading the niello while working under a flush of lukewarm water. I rarely use a grit coarser than #320, progressing rapidly through #400 and #600 to produce tonalities that are soft, matte, greyish and silky.

Power buffing speeds the process of finishing but presents a number of problems that require care and skill of execution. Niello will rise to a brilliant high polish with tripoli alone, which I view as sufficient final polish. I do not recommend the use of red rouge. Because power buffing will attack the niello faster than the silver or gold matrix, be gentle! Experiment with the range of finishing materials and tools at hand to determine which sequence works best for you.

Problem	Reason / or Suggested Solution
Pits in the finished work	Ideally all oxide scale is removed by fluxing and speedy firing during application. Extended heating and repeated remelting can often make the condition worse. With a small drill, cut into a pit to remove slag trapped there, then flux and remelt.
Flux is trapped below the surface.	The workpiece was not heated properly at the start, keeping the flux too firm for the molten niello to pass through it to the metal.
Silver lumps show up in the niello.	Insufficient sulfuration or serious overheating.
Dendritic patterns of an orderly structure in the silver grow and appear after the niello is polished.	Overheating during application for an extended period of time, which interferes with the sulfide bond.
The niello seems pale and metallic.	Similar to the condition described above, only more widespread and uniform.
After the pour, a core of grey silvery metal shows up in the rod, fully surrounded by what seems to be a successful niello.	Error in formula preparation, insufficient sulfuration or overheating during the process; the alloy was at too high a temperature when poured.
The niello rod appears grey and dull when cooled after the pour. This condition is best seen at the far end away from the point of the pour.	Insufficient sulfuration; too high a temperature at the pour; error in formula preparation; extended process time.
Upon examination, the underside of the rod has a granular coating of silver as a thin surface deposit.	This may not be a serious problem because it is a surface deposit that does not affect the center of the niello rod. Remove the film by careful scraping.
Upon polishing, the niello refuses to become clear, clean and brilliant.	Niello polishes and responds well to some compounds and not others. Try a different compound.
The niello does not flow into a truly Splendid Rod.	Not quite hot enough at the moment of pour, or perhaps a hesitation in pouring. The angle of the ingot mold might not be steep enough.

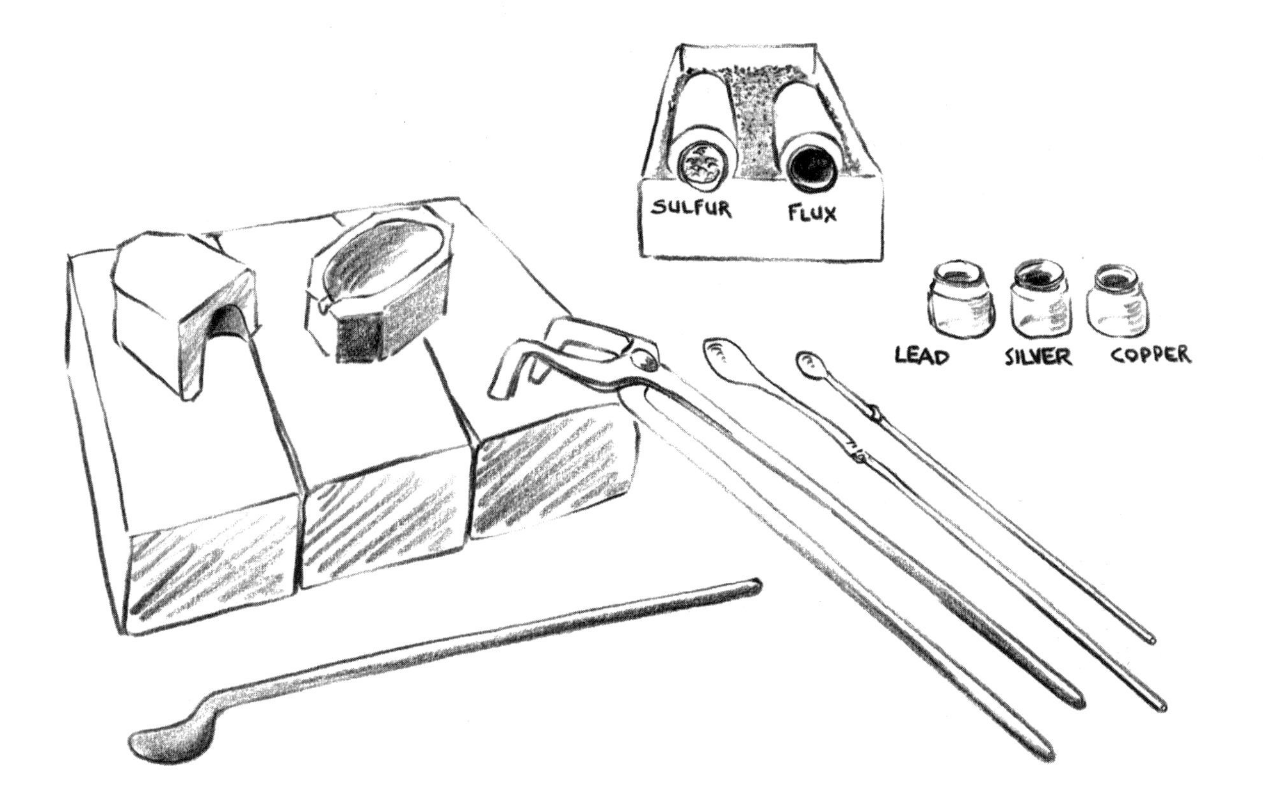

The Niellist's Workstation, ready to make up a batch. Care in preparation can make the difference between a rewarding success and a frustrating failure.

Phillip Fike is a professor at Wayne State University in Detroit and a leader of seminars around the US. He is almost singlehandedly responsible for the reintroduction of the fibula format into contemporary jewelrymaking, and a leading authority on niello.

Michael Good

Anticlastic Raising

For centuries, sheet metal forming primarily concerned synclastic forms; those in which the dominant curves all move in the same direction. This is a logical form to make (it's what happens when you cup your hands, for instance) and because of its utilitarian advantages it had immediate practical uses. An alternate raising method, in which the dominant axes move in opposite directions, was used only in limited applications such as the making of spouts or handle forms. Only in recent years have metalsmiths turned their attention to the technique called anticlastic raising and begun to explore the wide range of possibilities it opens.

I have narrowed my attention to shapes that can be made from a single sheet and it is these monoshells that I will illustrate here. For a fuller introduction to the vocabulary and techniques of anticlastic raising I recommend *Form Emphasis for Metalsmiths* by Heikki Seppä (1978, Kent State University Press).

Background

I first began working in thin sheet metal in 1968, with the idea of supporting myself by making jewelry. In my early work I used every technique I saw or read about, experimenting in a dozen different directions simultaneously. Inevitably, the resulting work lacked a unifying concept. After a disappointing foray into the marketplace, I melted down all my pieces and had the metal rolled out into a sheet of 27 gauge 14K gold. Ironically, this is the same thickness I still use in most of my work. The original designs from this period were simple flat shapes, cut out of this material and given a hammered surface for strength and looks. I next began to experiment with ways to manipulate the sheet to add strength as well as dimension.

Through trial and error I developed techniques of forming the sheet metal into hollow structures and built a line of jewelry on this theme. It was during this time that a fellow jeweler, seeing my work, told me about a master metalsmith and teacher, Heikki Seppä, whose manuscript on shell structures was about to be published. The following summer I attended a workshop taught by Heikki at the Haystack School in Maine, beginning a friendship and professional collaboration that continues to be one of the most important relationships of my career.

Synclastic
A metal forming process in which a flat sheet is shaped by compressing its edges and stretching its center. The principal curves of the resulting form develop at right angles to each other and move in the same direction. The sheet gradually assumes a domed form.

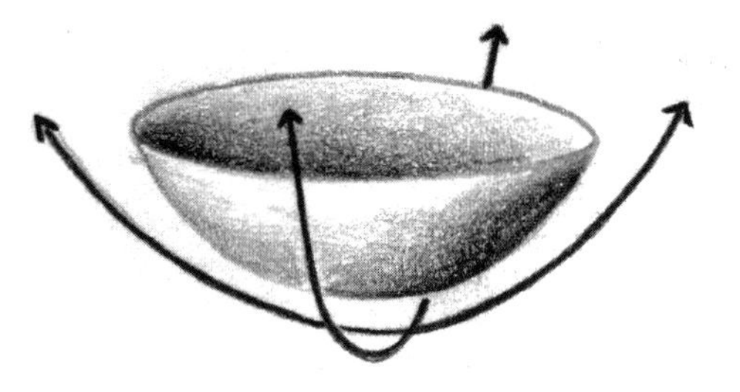

Anticlastic
A sheet metal process in which the center of a flat sheet is compressed while its edges are stretched. The resulting form develops two curves at right angles to each other moving in opposite directions.

Axial curve
A term used to describe the imaginary line that runs along the stake during anticlastic raising.

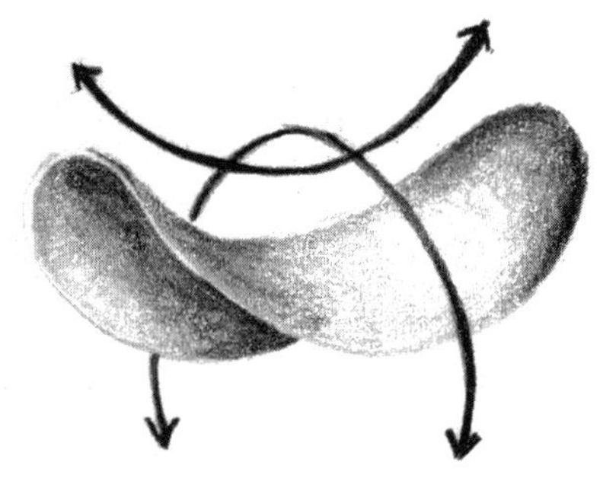

Generator curve
Another imaginary line, this one curving around the stake.

Sinusoidal stake
This is a tool, generally of steel, wood or plastic, that is used to support the metal throughout the forming process. These can be purchased or made in the studio. The word derives from the Latin term for wavy or winding.

Cross peen
This is a mallet or hammer with a wedge-shaped face that runs perpendicular to the handle.

TOOLS AND MATERIALS

Anticlastic forming is done on sinusoidal stakes made of wood, hard plastic or steel. Generally a wood or plastic cross peen mallet is used when working on a steel stake, while a steel hammer is used against a wood or plastic stake. A steel stake allows for swift deforming of the metal, but the process is a little harder to control. Because of this it is not unusual that the workpiece will require additional planishing in its final stages. In other words, time saved in fast shaping is lost in extra planishing. Conversely, use of a wooden or plastic sinusoidal stake allows for slower but more controlled anticlasting, in some cases eliminating the need for planishing altogether.

Tools are chosen in a size that will be appropriate to the shape of a given pattern. The dimensions of the stake determine the limits to which a sheet can be formed. The cross section of the stake at any given point determines the maximum axial curve. Similarly, each sinusoidal arc determines the maximum generator curve. In the case of smaller patterns, the width of the starting blank should not exceed the distance between the two crests of the sinusoidal stake. In every case, the arc of the cross peen hammer or mallet should be slightly tighter than the corresponding curve on the stake. These relationships are illustrated in the following drawings; see the drawing at the top of page 35.

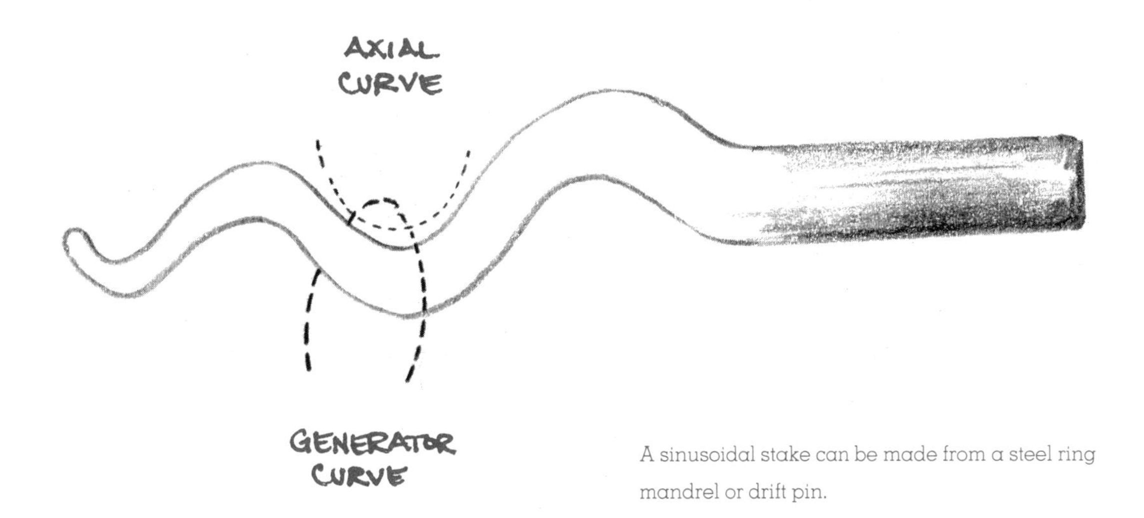

A sinusoidal stake can be made from a steel ring mandrel or drift pin.

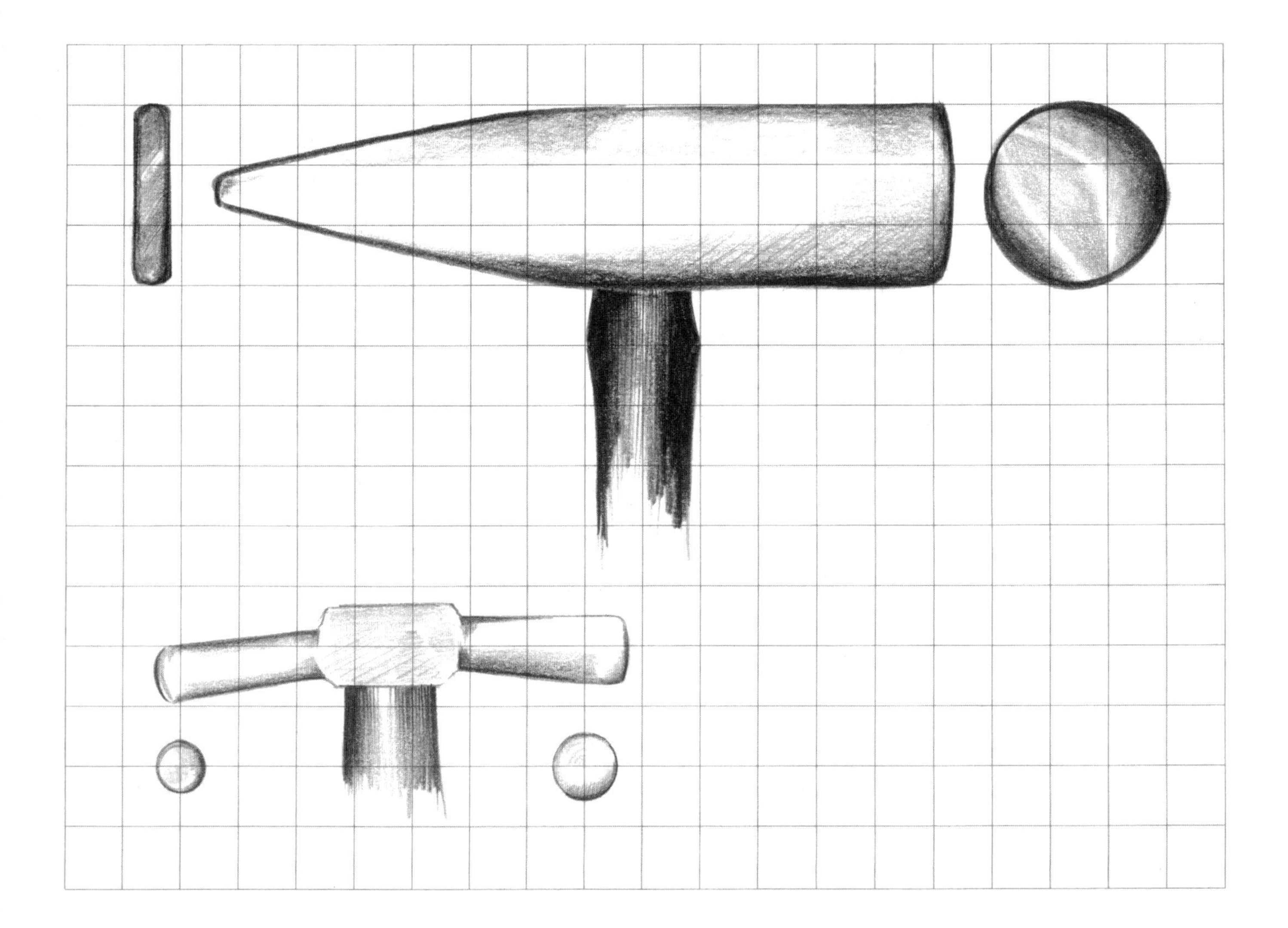

Each grid square represents ½" – enlarge these illustrations 133% to create templates.

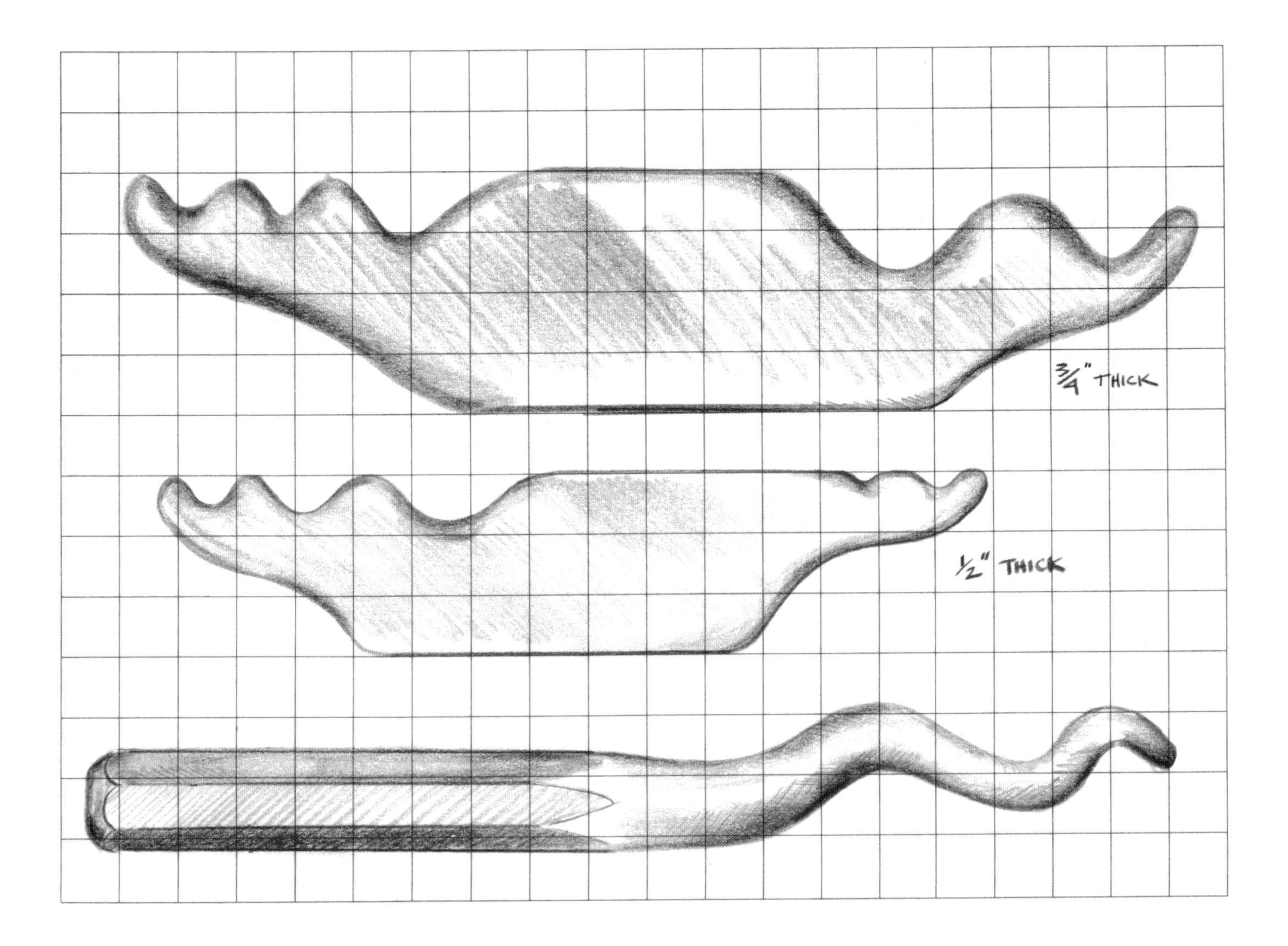

The mallet at the left and the two upper forms on this page are made from a dense plastic such as Delrin or nylon. The hammer and the sinusoidal stake are steel.

FORMING PROCEDURE

For the purpose of instruction, I'll describe the process of making a spiculum like the one shown here. This shape requires all the various steps of forming, closing and finishing an unsoldered, anticlastic monoshell. Once you've mastered this, you'll be able to experiment with other forms and techniques. In the interest of clarity I have provided an exact pattern shape, which should be cut from 27 gauge metal. This is offered only as a resource, with the expectation that the more adventuresome will want to work from an original form. Whatever it is, it's a good idea to trace the shape of your starting piece into a notebook so you'll have a record that will allow you to duplicate the piece if it's right, or avoid making the same mistakes if its wrong.

Cut out the shape and file the edges until they are smooth and regular. Smoothness is important because jagged edges might cut your hand as you work and regularity is important in the looks of the final piece. Symmetry is not necessary for anticlasting, but we'll start with a symmetrical shape for this example. Anneal and dry the piece and you're ready to begin.

The technical name for this form is an open seam anticlastic spiculum. This is the piece that is being made throughout the article.

This is an actual size pattern for the spiculum.

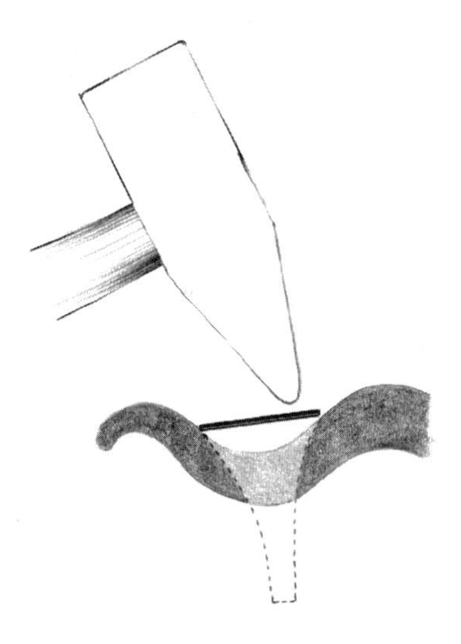

The strip will rest on the stake
in such a way that there is a void beneath it.
Begin striking at the upper edge.

Secure a sinusoidal stake in a vise, and bend the annealed metal sheet around the sinusoid so that the edges are supported at each end of the curve. Hold the pattern firmly at each end so it doesn't straighten out when struck. Direct the hammer blows to just below the point where the stake and pattern meet, so that a small amount of metal is curled against the stake as shown. Rotate the piece across the stake so that each successive blow slightly overlaps the previous one until an even furrow is formed along the edge. Start at the tip of the form and always work toward the center.

Unless done intentionally for special effect, care should be taken to strike the metal at right angles to the stake. If hit obliquely, the pattern will twist either to the left or to the right depending on the angle at which it was struck. If the metal pattern has a tapered shape like this example, more forming will be required at the wider sections of the pattern. Strike as many courses as necessary there, blending them into the taper to insure that the form is uniformly curved.

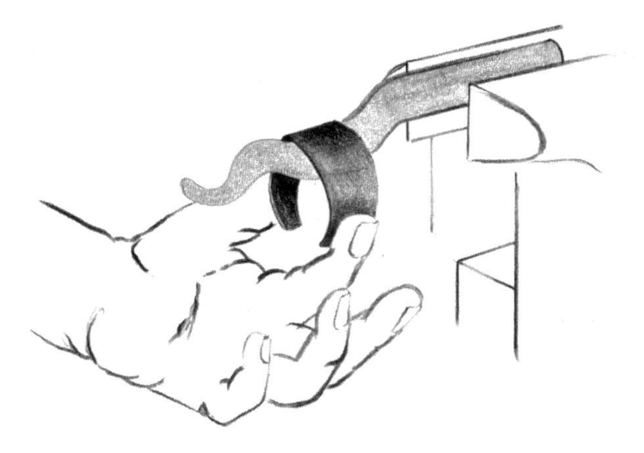

It's important to hold the ends of the strip
together as you hammer.

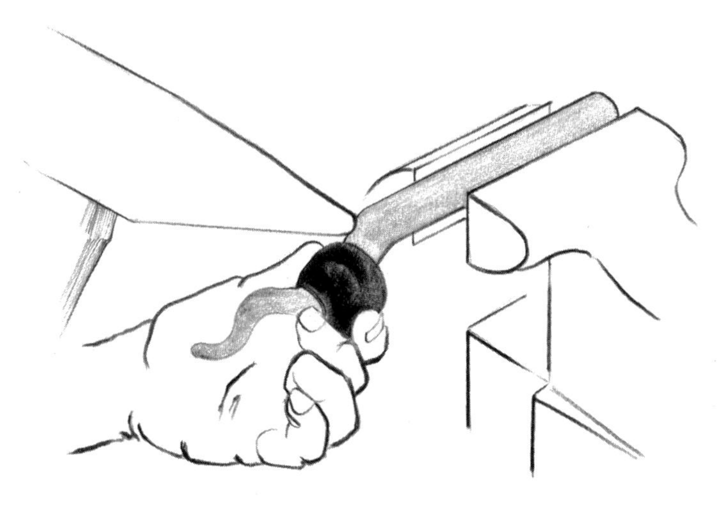

The spiculum is rotated around the stake
as you continue to strike with firm blows at the
same spot on the stake.

Turn the piece 180° on the stake so the unformed edge is now resting on the end of the stake curve closest to the vise and repeat the furrowing process in the same way. This completes the first hammering pass over the pattern.

Place the furrowed piece on the sinusoidal stake so there is a "hollow" beneath the form, like the one shown in the middle drawing. Direct the next series of blows just below the point of contact, rotating the piece a little so that each blow slightly overlaps the previous one. Strike a series of overlapping blows as you slowly rotate the workpiece, creating a second furrow just inside the first.

Turn the piece 180° on the stake and repeat the process on the opposite edge until the second pass is complete. As with the first raising passes, always work toward the center. As the metal contracts it will tend to form a lump along the axis. If you try to move the metal too quickly, the lump will get out of hand and cause an unwanted fold in the developing pattern. If this starts to happen, anneal the sheet and resume anticlasting with lighter blows that overlap more.

Continue raising in this manner, moving systematically toward the center until the first anticlasting course is complete. Remember to overlap blows with each succeeding pass. The work-hardened form is then carefully annealed to whatever temperature is appropriate for the alloy being raised.

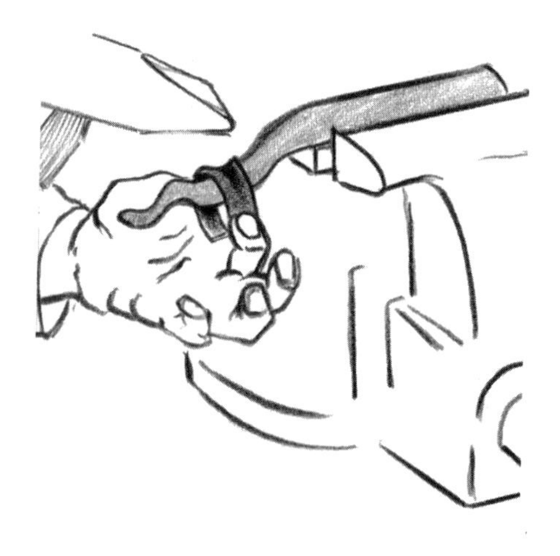

After traversing the entire length of the spiculum on one edge, slide it off the stake and turn it around, reposition and repeat the process.

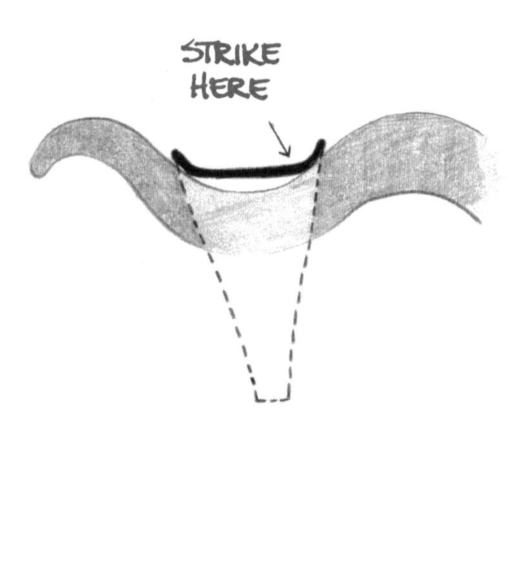

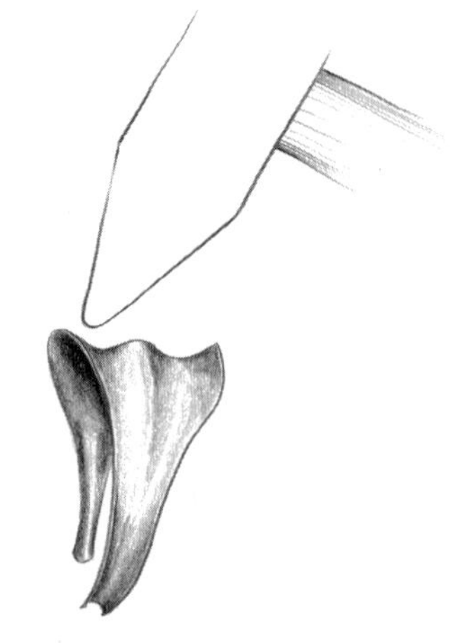

As you work along the second edge, a bulge will form down the center of the strip. This must be pressed down and made smooth before the spiculum is developed further.

To begin the second course the form is bent around the sinusoidal stake in a process similar to the first step. When the axial curve is tightened the generator curve opens, so it will appear that your nicely closed form has opened. Well, it has opened up, but don't be discouraged! If the limits of the first curve have been reached, it might be necessary to move to a tighter curve on the stake. Except in the case of exceptional hammership, it usually takes two courses of hammering on the first portion of the stake before this becomes necessary. You can continue to work on the steel stake or, as shown, you can switch to the plastic stake. Experiment with both to determine which you prefer.

This process will continue until the desired form has been obtained. Strike with a cross peen on the inside of the trough as it develops, always holding the ends of the form together to minimize their natural tendency to pull outward. Position the workpiece on the stake so there is a hollow or void beneath the metal where you intend to hit it. Remember that the sinusoidal stake can be turned over to reveal a series of curves with different radii.

The anticlasted strip has several interesting characteristics. While it has great structural strength, it nonetheless moves freely about its axis and can be easily twisted in both directions. By "screwing" or "unscrewing" the form, the axial curve will open out and the generator will close.

Continue alternate working of both edges, moving from the outside toward the center line.

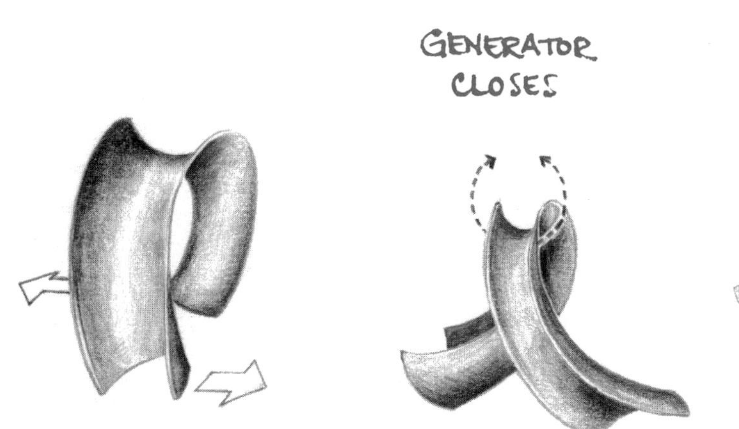

The orientation of the ends of the strip will have a lot to do with the resulting form.

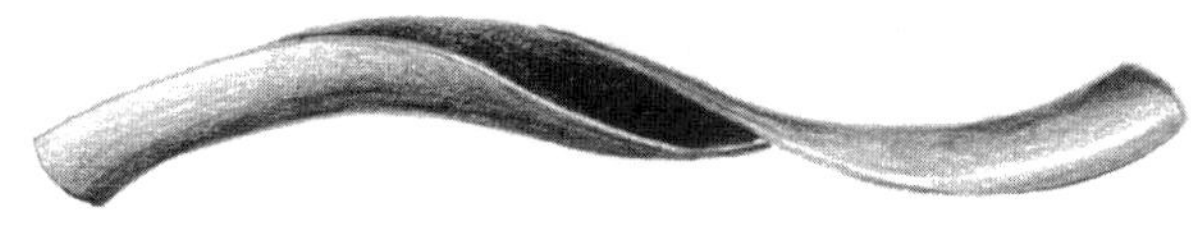

By systematically bending and twisting, the sinusoidal form can be turned in just about any direction. The concave side of the anticlast will always remain on the outside of the curve, adding strength and definition to the forms. When the anticlast form is bent along its axis, that axis is tightened and the generator curve will tend to straighten.

When a particular direction has been chosen, usually after the first course of anti-clasting, the workpiece is returned to the sinusoidal stake for further anticlasting, making sure to maintain the form until the desired final shape is obtained. If the form starts to grow in an unwanted direction, grip it in the hands and bend it back into shape. It is important to pay attention to the form as it develops so you will see any unwanted evolution soon enough to stop it. Things can quickly get out of hand!

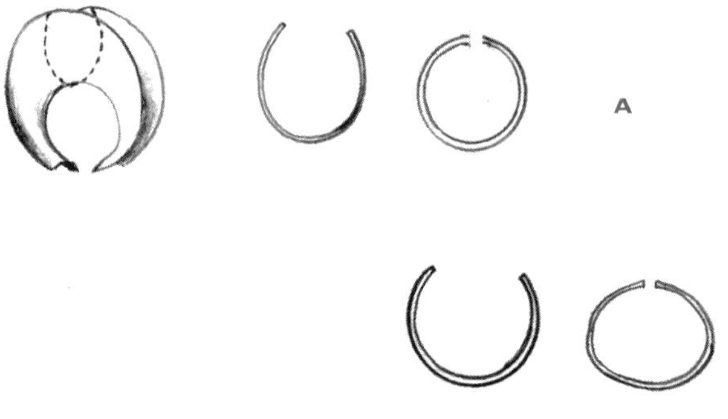

A

B

The cross section of the midpoint should be an oval in preparation for closing the form (A). If the cross section is round at this stage (B), the spiculum will be oval by the time the edges meet.

If there is a most common problem as people learn anticlastic raising, it is a tendency to move up the form too quickly. I cannot over emphasize the importance of achieving a smooth, regular curvature in the lower half of the form before trying to bring in the sides. There is simply no shortcut to the smooth lines and look of a well made anticlast, and no hiding a sloppy job either. Take your time and get it right before you try to complete the form.

Planishing, by definition, involves a minor thinning of metal to achieve a planar surface. In my case, this requires a precisely controlled pinch of the workpiece between a plastic stake and a steel hammer. It is obvious that both the hammer face and the stake must be painstakingly smoothed to achieve the intended result.

When it's ready to begin the closing phase, the form in cross section will be oval with the opening directly in the center of the top. As the upper walls are brought in they will form a circular section in the final form. If the form appeared round at this stage it would become a flattened oval when closed.

Clamp a small, polished ring mandrel or drift pin into a vise, extending horizontally. If you'd like to lay a memorable trap for passers by, leave this sticking out when you step away from the vise for a few minutes. If not, remember to remove it! Hold the form along the side of the tool, matching the curve of the mandrel to the curve of the form. Because it is making contact at only one spot, the position of the mandrel is important. The correct location is shown on the next page, exactly at the point where the curve should next be brought in. In practice this is more a matter of feel than visual reference. Use the larger diameter of the mandrel for the larger areas of the form and move to the mandrel tip as you rotate the form to its narrower ends. Continue in this way (proceeding slowly!) until the form is closed with a smooth-edged uniform seam that runs symmetrically along the spiculum.

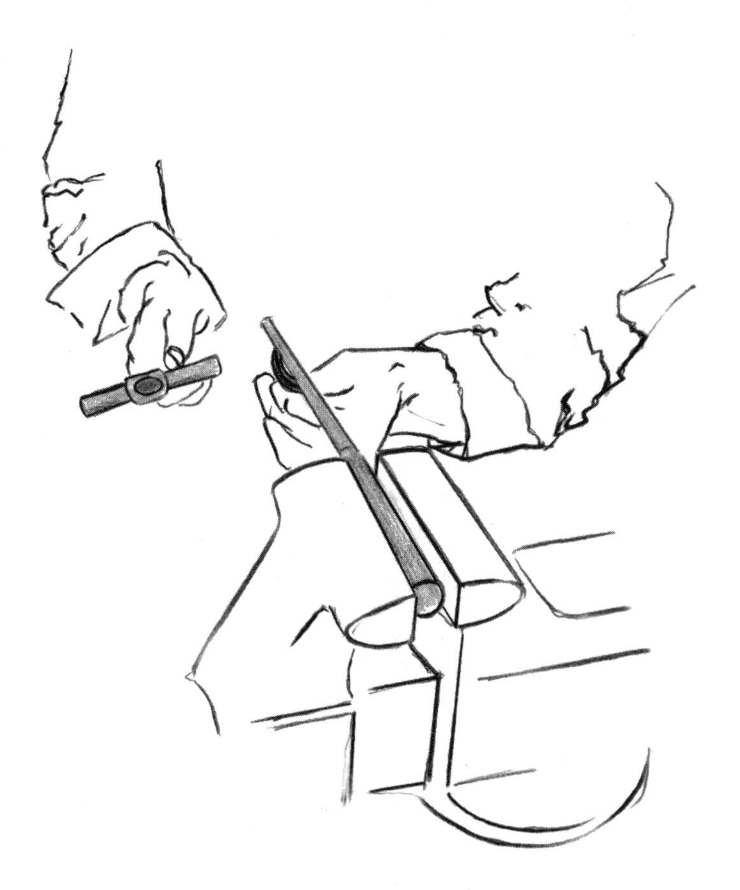

Working against a standard ring mandrel, pinch the sides of the spiculum to close it.

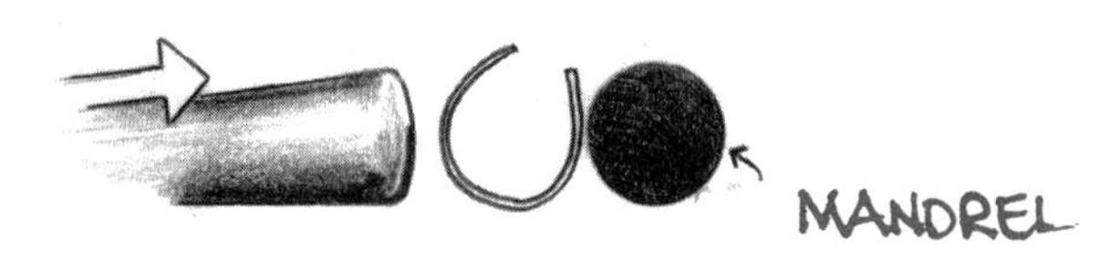

The curved faces presented by the hammer and the mandrel make a minimal contact with the spiculum. A delicate feel is needed to locate the ideal tangent spot that will close the form.

It is worth noting that at this step you have switched from anticlasting to synclasting, or the process we usually call raising. The metal sheet is being pushed in and up along axis lines that both curve in the same direction. Because of the impossibility of working over a rigid stake, the metal is worked against its own rigidness and depends on the form being work hardened. If a stout blow was delivered to an annealed spiculum at this stage it would collapse. The action might best be described as a tentative "feeling" stroke of the hammer, with care taken to insure that you are exactly opposite the contact point where the form lays against the mandrel.

Summary

Part of the beauty of the anticlastic method is its ability to achieve a great deal of structural strength with a small amount of material. This means precious or heavy metals can be used in ways that might be impossible with other techniques. The principles of the process are adaptable for large and small forms, so the scale of your experiments will be limited only by the tools available and your ability to handle the material. The combination of anticlastic and synclastic raising creates unlimited potential for exciting new monoshell structures.

Michael Good is a designer and manufacturer of gold and platinum jewlery for sale in the US and Europe. When not traveling, he lives in Camden, Maine.

Susan Kinglsey

Nonconforming Dies

Die forming is an an ancient process and has been used throughout history by many diverse cultures. Etruscan and Greek goldsmiths, for example, used dies to create multiple units for earrings and necklaces. They carved designs into bronze plates and hammered thin gold sheets into them through thick sheets of lead. Using the same idea, I replace the bronze with a carved acrylic die, substitute urethane for the lead, and deliver the force in a press with hydraulic pressure. My use of nonconforming dies is a continuation of an age-old practice with nineteenth century technology and twentieth century materials. Surprisingly, hydraulic presses very similar to those described here were in use before 1850.

I began working with die forming in 1978, shortly after Marc Paisin described the process in *Goldsmith's Journal* (VIII, #6, December 1977). His research was based on the Masonite® die technique developed by Richard Thomas at Cranbrook Academy and on work done by Ruth Girard at the University of California at Berkeley. I have an artist's curiosity and a metalsmith's love of tools and process. Most of what I know about die forming was discovered by trying out "what ifs" and then figuring out what to do with the result. Teaching workshops in die forming has given me the opportunity to learn from the successes and failures of other metalsmiths. In 1990 I began working with Lee Marshall, an engineer and inventor who had designed a small hydraulic press. He formed a company called Bonny Doon to develop and manufacture tools for die forming, and has opened the possibilities of this exciting technique for a larger audience. In a wonderful snowballing effect, this activity has lead to more inventiveness and further discovery.

INTRODUCTION

A die does the same thing as a hammer, stake, dapping punch or repoussé tool, but replaces muscle power and repetitive blows with the long slow squeeze of hydraulic force. Die forming is a direct, efficient and fast process that challenges our assumption that metalworking is slow and demanding. The die forming process allows metalsmiths to make full use of the plastic qualities of metal with directness and efficiency.

Die forming permits the reproduction of many standardized objects, but this is only one of its many possibilities. It is important to remember that dies are tools, not molds that simply reproduce identical objects. Dies shape and mark work in the same ways that a specific stake or hammer leaves its character on a piece, and the end result depends upon the expertise of the user. Die forming can add dimension and volume to work, and because formed metal is structurally stronger than flat sheet, it enables the use of thinner metal. This allows jewelry to be made lighter in weight, both increasing its comfort and lowering its production cost in precious metal. Forming with dies is less stressful to the metal than hammering because force is evenly applied over the whole surface. This makes it possible to form mokumé or married metals with less danger of splitting. Similarly, roll printed and etched metal can be formed without damaging the surface. In every case the formed metal is smooth and free of hammer marks.

The processes described here are recommended for nonferrous metals. Copper, sterling, fine silver, pewter and aluminum are easily formed with dies. Yellow golds and brass work well, as does titanium and niobium. Stiffer metals such as nickel silver and bronze are not recommended.

DIES

Standard hole punches can be used in a hydraulic press and various tooling is available for doming, dapping, stamping, and bending. Blanking dies, such as RT dies can also be used in the press.

Conforming dies consist of two corresponding rigid male and female parts between which metal is formed. They are essential in some cases, when complex shapes are required. Conforming dies for use in a hydraulic press can be "home-made" using another twentieth century material, Devcon® steel filled epoxy. Detailed information concerning conforming dies, and the use of this material may be found in Marc Paisin's article in *Goldsmiths Journal*, listed above, in my article "Hydraulic Die Forming for the Artist/Metalsmith," *Metalsmith*, Summer 1985, or my book, *Hydraulic Die Forming for Artists and Metalsmiths.*

Nonconforming dies used with urethane offer many exciting possibilities for creative work. Because only one part is needed, these dies are easier and less expensive to make than conforming dies. Also, they allow a greater degree of versatility in their use. The next section describes the equipment needed for die forming, and is followed by instructions on the making and using of three types of nonconforming dies: punches, matrix dies and embossing dies.

Press Frames

Many of the "first generation" welded steel press frames from the 70s are still in service. These frames are constructed of U or L channel steel with a "free" platen, or middle plate that rests on the jack within a section of pipe welded on the bottom. It is very important that the platens are flat and parallel. The addition of return springs makes them easier to use.

In 1979, Robin Casady designed a "bolt together" press after consulting with a machinist to determine the specifications to withstand 20 tons of stress. One advantage of this press (beyond the fact that you don't have to be a welder to build it) is that it is adjustable to various heights, allowing the possibility of doing larger work. The cost of this press will vary depending on your source of supply and on how much of the work you do yourself. Because of the amount of steel that is necessary for the plates, and the cost of having the plates drilled, it could be quite expensive.

The Bonny Doon press consists of a welded frame of 2 inch square steel tubing and 6 x 6 x 1 inch steel plates. Heavy duty springs concealed within the frame provide both a self-leveling feature and "return." This is a compact, safe and efficient design for studio work.

The inexpensive shop presses available through discount tool suppliers are not recommended. They are unnecessarily large, often underpowered, require too many modifications, do not have pressure gauges, and tend to break easily.

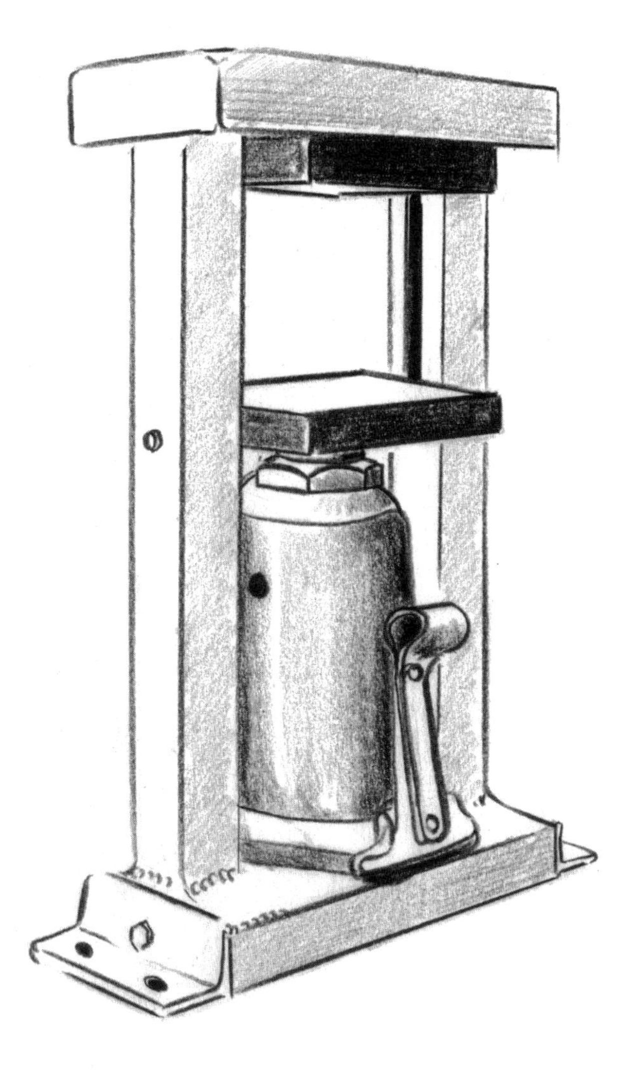

Bonny Doon Press with bottle jack

POWER

The basic power unit for the die forming press is a standard 20 ton hydraulic jack (sometimes called a bottle jack) available at automotive and industrial supply houses. Ten ton jacks are adequate for some types of dies, but nonconforming dies require a twenty ton capacity. The addition of a pressure gauge makes it possible to measure the force being applied, permitting control and consistency, and is highly recommended.

An alternate hydraulic system is a separate two stage pump and cylinder connected by a hydraulic hose and gauge. This can be put together from components available from industrial or commercial equipment suppliers. This arrangement is somewhat more expensive, but the two stage pumps make it easier to operate.

A third alternative is the use of compressed air to power the pump of the hydraulic jack. This system is recommended for production work, and for anyone who requires a process which is less physically demanding. The air pressure can be set for accurate reproduction, and the cycle time is faster than hand pumping.

URETHANE

Urethane is a tough rubber-like substance. Used in conjunction with nonconforming dies it becomes, under pressure, the other half of the die. Unlike rubber, which compresses and degenerates with use, urethane flows under pressure, distributing its force evenly over a large area without percussive impact. After use it returns to its original shape and can be reused thousands of times. The ragged appearance of the material after extended use does not affect its performance. Sheets or slabs of surplus rubber can be used with nonconforming dies, but the results are unpredictable and less dramatic, and rubber tends to degenerate quickly.

I find it useful to have a variety of urethane pads and blocks available in different sizes, thicknesses and durometers. Hardness is measured in durometers, and urethane is available in 95, which is the hardest and 80, which is softer. In general the 95 is used for forming, while the 80 is recommended for embossing and "intensifying" some forming operations.

Because it "flows" under pressure, the urethane is usually contained in a strong walled vessel when used with punches so that it flows up and around the punch. A section of ¼ inch steel pipe filled with urethane to within a half inch of its top edge is a typical arrangement. For general use, these containers are bottomless. To achieve the greatest force, resulting in the maximum detail, use a container with a bottom. Containers are usually used with 95 durometer, but 80 durometer urethane is useful when extreme detail is required.

Urethane pads are made in 80 and 95 durometer and in thicknesses from 1/16 inch to 1 inch. They are used when the die will not fit the contained block. Matrix dies are usually used with pads. The thickness used depends upon the depth and size of the die. Larger dies, in which deeper forming is desired, require thicker pads. The thinnest pads are used for embossing.

The two part pourable, self-curing urethane Flexane®, has been recommended in the past for making these pads and blocks. Because it has recently been identified as a hazardous material, it can no longer be recommended. The danger occurs while it is being mixed and cured: air purifying respirators do not provide adequate protection. Flexane® pads that are already made can still be used safely. The material should always be cut with scissors or a sharp knife.

BECAUSE DANGEROUS FUMES ARE RELEASED, NEVER HEAT, GRIND, BURN, SAW OR SAND ANY FORM OF URETHANE.

PUNCHES

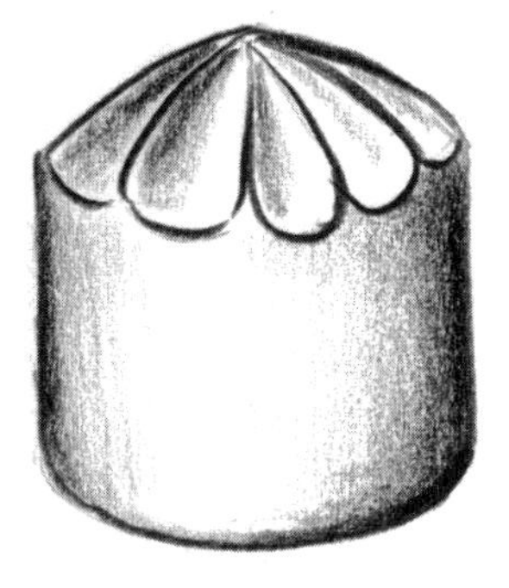

A typical punch carved from acrylic.

A punch (sometimes called a "male" die) is a tool that is driven against and into metal to form it. It is usually a "positive" of the desired shape, as shown below. Contained urethane provides the other part of the die.

Cast acrylic (Acrylite,® Plexiglas® or Lucite®) rod makes a surprisingly good die. Acrylic is widely available, inexpensive, easy to work, and has a compressive strength of around 1800 PSI (pounds per square inch). In cases where crisp edges or long wear are anticipated, I recommend the use of Delrin, a stronger and slightly more expensive material. Whatever the material, the ends of the rod must be cut at 90° to the axis, with parallel faces. I find one inch lengths to be a handy size. It is difficult to cut sections accurately, so I recommend having it cut by the supplier.

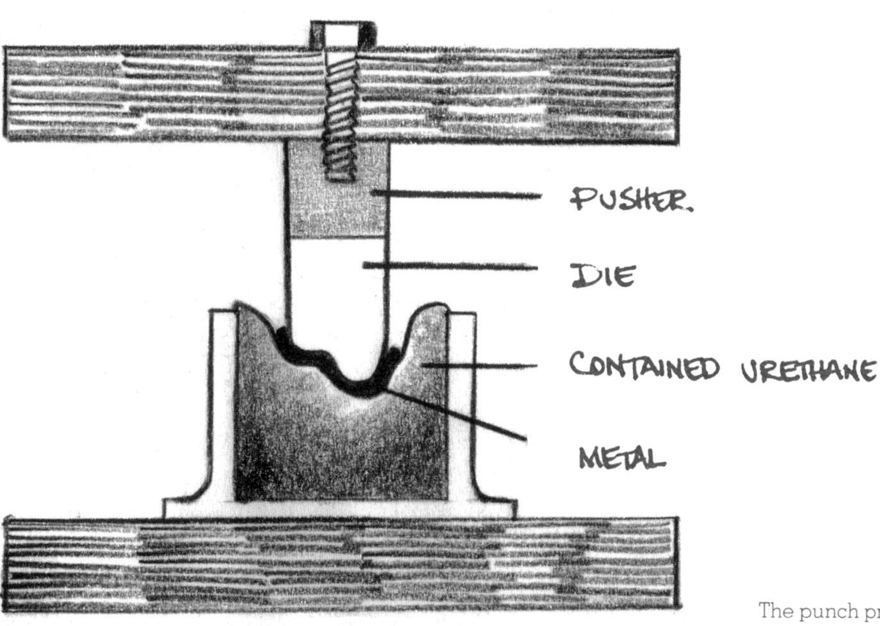

The punch presses into the urethane pad, which flows up and around it. The urethane is contained in a cylinder with a welded bottom. Note that the punch is bolted to the top platen of the press.

Acrylic can be shaped with coarse files or carved with burrs in a flexible shaft machine. Wear a dust mask and goggles when working with acrylic, and avoid grinding at high speed. If the acrylic becomes "gummy" or emits a sharp plastic odor, you are overheating, and should slow down and apply less pressure. The decomposition products released during heating can be hazardous. Vacuum up all dust as soon as you are finished.

As you carve a punch, first rough in the shape. Overly deep cuts and undercuts will break down the die. Steep angles may tear the metal before it can form. Make some impressions of the shape as you go along before spending much time on the surface finishing or detail. Punches are tools for forming and contour should be the first consideration. Texture can also be part of the design, but will be effective only when a thin metal (28 gauge or thinner) is being used.

Other suitable materials for punches are end grain hardwood, nickel, brass, aluminum or steel. Dies can be carved, constructed or cast. The choice of die material will depend on the size and complexity of the die, the material being formed, the number of reproductions needed and the materials at hand. Do not use any material that has the possibility of being brittle, or anything "long-shafted." The one inch lengths of material suggested for acrylic dies are also recommended for other materials. Because they are pushed down into the steel-walled container during use, the die would be enclosed in the unlikely event of breakage.

To use the punch, select a contained urethane block which is close to the size of the diameter of the die. The punch can be secured to the top of the press in several ways. It can be bolted directly to the top platen, or taped in place with double sided Scotch® tape. Because the urethane does not fill the container, a spacer or "pusher" may be necessary to push the punch down into the urethane. I find it most convenient to make my punches of one inch lengths of material and tape them to a pusher which is bolted to the top platen.

Always work with annealed metal. The metal blank should be smaller than the die. If a larger piece of metal is used, it will wrap itself around the die (like a bottle cap) and may be difficult to remove. It could also prevent the metal from being drawn into the die, causing a poor impression. The thickness of the metal is what determines how much surface detail will be achieved: the thinner the metal, the more detail. To maximize detail, you can do several things. One is to anneal and do repeated pressings. The formed blank can be held in place on the die with masking tape. Another way to maximize the impression is to use a contained block with a bottom. 95 durometer urethane is usually used in the container, but containers of 80 (with a bottom) are most effective for obtaining maximum detail. Experiment and write down the procedure for each sample so you will be able to reproduce your results.

When using a punch in the press, you can watch what is happening. The punch will go down into the container, burying itself and the blank in the urethane. You can push the punch as much as two thirds of the way down into the container (if it has a bottom). If it is a bottomless container, stop when the urethane starts to ooze out the bottom, because you have reached maximum pressure within the container.

Matrix Dies

A matrix die.

These can be made of plastic, steel or aluminum.

A matrix die could also be called a silhouette die, as it is simply a block of material with the outline of a shape removed, leaving its silhouette. A matrix die is a female die; a die within which a form is shaped. Urethane, when used with a matrix die, becomes the punch.

Matrix dies are similar in concept to the Masonite® dies that have been used by metalsmiths for many years. In this case, however, it is not necessary to secure the metal to the die, or to use a hammer or dap to form the metal. The press "clamps" the blank against the die, and the urethane pushes and stretches the metal smoothly and evenly into the opening, without any marks. The result is a gentle pillow-like form with a flat flange, and concise outline.

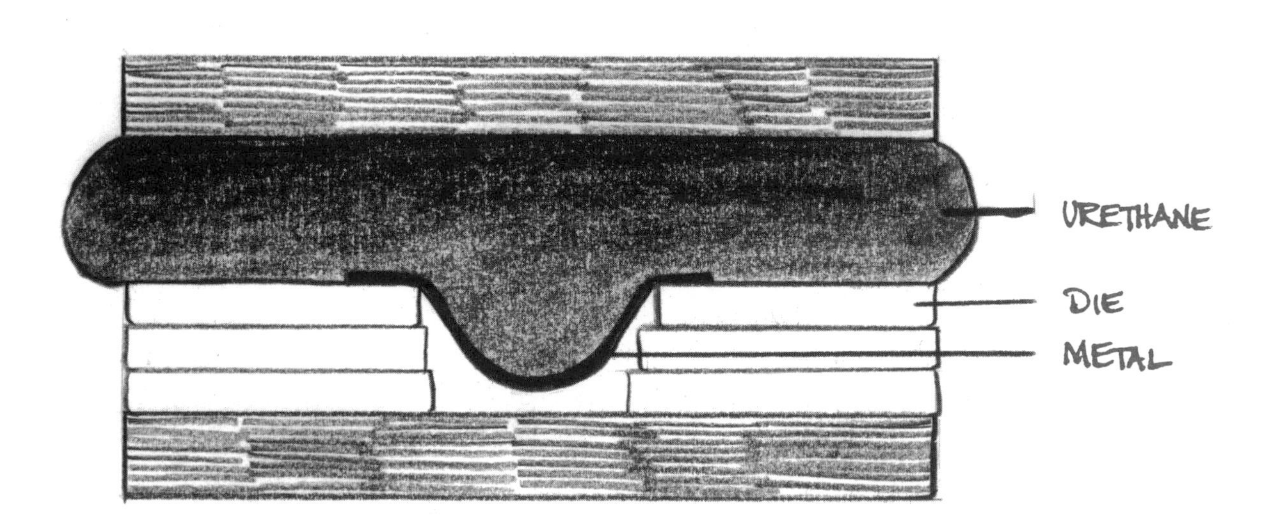

In the matrix die the urethane takes over the role of a punch, pressing the metal into the die.
Note the use of layers to increase the thickness (depth) of the die.

Masonite,® wood and aluminum have been used to make these dies, but I prefer acrylic sheet because it is easy to cut, inexpensive, and long lasting. The addition of a brass or steel face plate makes a die that can be used hundreds of times. Acrylic is available in a range of thicknesses, but I find that ¼ inch is the most useful, as it is less expensive, easy to cut, and can be layered to achieve a range of thicker dies. Large sheets can be cut into more useful sizes with a table saw, and some plastics suppliers will do this for you. It is also possible to buy scrap pieces by the pound. If you can afford the initial investment, you might lay in a supply of squares of useful sizes. It's great to be able to simply pull a blank off the shelf and get right to work.

To determine the best thickness for your die, measure the widest part of the silhouette you are cutting. The wider your form, the deeper the relief you can press and therefore the thicker you should make the die. The following chart gives a general idea of the sizes involved.

Width of silhouette	*Thickness of die*
1"	¼"
1½"	⅜"
2"	½"
2½"	¾"

Center your design on the die, and leave a border of at least ¾ inch around the silhouette. Without this margin, the die is likely to break, especially if there are "points," as in the illustrated example. A margin of an inch or more is even better.

I cut acrylic sheet with a spiral sawblade set into a jewelers sawframe. Because these blades do not cut a clean edge and are difficult to control, saw near rather than on the line and then file to the exact shape. It's important that the top edge of this be cut cleanly and that the wall be vertical with no undercuts, but the finish on the interior wall is not important.

Generally I leave about an inch of solid material all around a die cutout. Avoid placing points near an edge as at the lower tip of the heart.

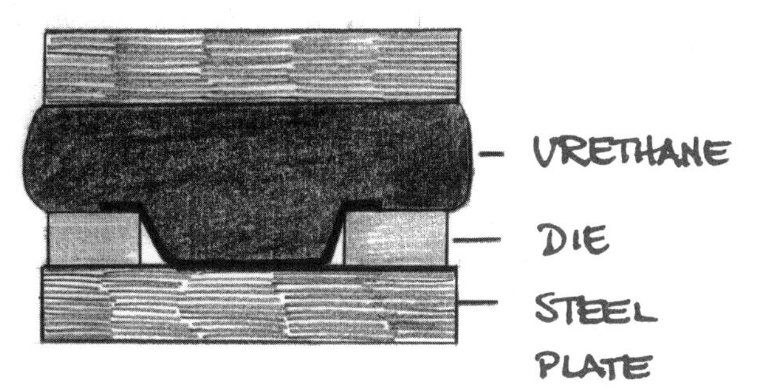

"Bottoming Out."
The solution is a thicker matrix die.

If you want to build up the die from several layers of ¼ inch acrylic, simply cut out the first silhouette, then trace it onto the next sheet with a scribe. A perfect fit is not vital, but undercuts will cause the die to break. Acrylic has great compressive strength, but it is brittle and will crack and break when it is not fully supported. The layers can be permanently glued together with acrylic cement, but I've found that double sided Scotch® tape works just as well. Of course layers can be added at any time, so remember this trick if a die "bottoms out" by pressing the metal against the bottom of the press.

One of the uses of die forming is to create matching "halves" that can be joined together to create a hollow object, such as a bead or small container. When the silhouette is perfectly symmetrical, the edges of two formed pieces put back to back will have matching edges, but if the design is asymmetrical, they will be mirror images. If you cut the die on a scroll saw or with a router, however, you can make a die with perfectly vertical "walls." The silhouettes on each surface must be perfectly matched mirror images, so you can press one sheet of metal into one side, flip the die and press into the other side, and the "halves" will fit together like clamshells.

Another way to make a double sided die is to make matching face plates. Glue two pieces of 16 gauge brass or 1⁄32 inch steel together, and carefully saw out the design. File so that the edges are 90° and then separate the plates. You can then make a single two-sided die or a pair of dies by using the surfaces that were stuck together as face plates on an acrylic die with opposite images. The metal plates can be glued in place or held with double sided tape. Again, make sure there are no undercuts in the acrylic.

Brass facing plates that have been sawn and filed together insure that the edges of the two sided die are accurate.

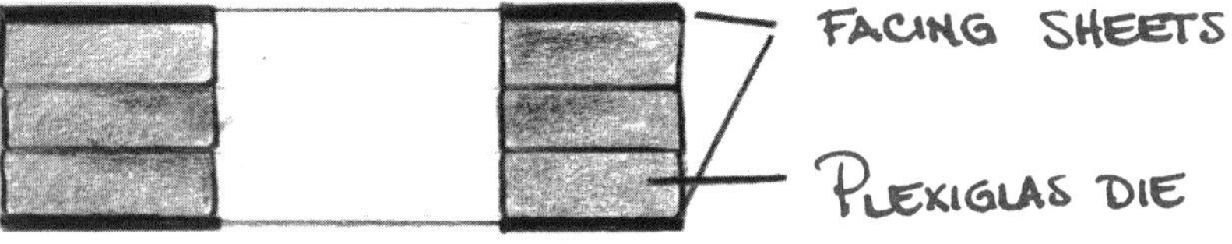

Metal from 18 to 28 gauge can be formed in matrix dies. The thickness you choose depends on the degree to which the metal will be stretched. Again, mark each sample with notes on the procedure so you have a record. Deep dies (with wider silhouettes) require heavier gauge metal than shallow (smaller) ones. The eventual use of the die formed part should also be considered in determining metal thickness. Because a flange adds structural strength to the form, a thinner metal can be used when the flange is to be retained in the finished piece. If matching parts are to be joined together without the flange, use a metal that is thick enough to allow a solderable edge.

The blanks for a matrix die should be ½ to ¾ inch larger than the silhouette. This will create a flange of ¼ to ⅜ of an inch all around. Matrix dies are generally used with the 95 durometer urethane pads. The thickness of the pad depends upon the width and depth of the silhouette, and should be in proportion. The one inch pad should be used for forming the larger, deeper dies, and the ¼ inch pad for the smaller, shallower ones.

The 80 durometer pads are also used with matrix dies, but you must be cautious. Because the softer urethane flows more easily, the forming process will happen faster. Metal can be drawn into the die before the flange is clamped firmly into place. This causes the edges of the die to be stressed by the drag of the metal, and damaged. To avoid this, I suggest using the 95 durometer pads to get the form started, and then switching to the 80. The effect of the softer urethane is a slightly fuller shape.

When using matrix dies, you should expect to anneal and press several times, stretching the metal slowly into the die. As with other metal forming processes, it must progress in stages. The combination of twenty tons of power and 80 to 95 durometer urethane are an awesome force, which you must learn to control. Begin at lower pressures, anneal frequently, and keep a record of the pressure used. Metal forms quickly, but it also work hardens quickly.

It doesn't matter which way the "sandwich," (urethane, metal, and die), is put into the press, as long as the metal is between the urethane and the die. It takes some experimentation to arrive at the desired result. I generally begin by making tests of a new die with copper. Try using various metal thicknesses, different pressures, different thicknesses and durometers of urethane, and vary the number of pressings and annealings. Write on each sample what you did to achieve it, so that you have a record.

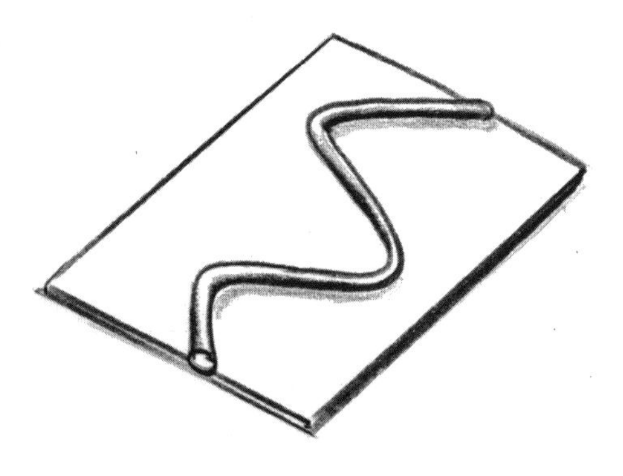

The two kinds of embossing dies:
In the first (above) a brass wire is soldered to a brass sheet. In the second (right) a line is carved into a plastic sheet.

Embossing Dies

The two types of dies just described are used to create forms. By contrast, embossing dies produce a shallow relief in thin metal. Embossing is different from stamping and roll printing in that the metal is formed from both sides.

The first drawing shows how stamping and roll printing thin the sheet. Embossing, by contrast, pushes the metal up to create a relief without substantially thinning it.

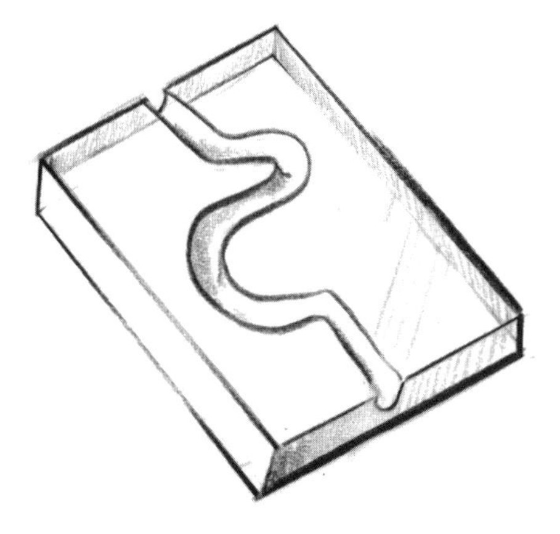

Embossing dies can be assembled from found objects. Washers, lock nuts, metal screen and small gears can be used as dies. Various textured plastics can also be used, such as the window pane material that is used for lighting fixtures or bathroom windows. "Organic" materials such as leaves simply compress and "natural" materials such as shells break down and are unsuitable. Do not attempt to press rocks because they can shatter.

Embossing dies can be made in a variety of ways. Wire can be soldered to sheet to make a simple punch type embossing die. Matrix type embossing dies can be made by piercing a pattern into a sheet of 16 gauge brass. A shallow relief can be carved with burrs into acrylic. In order to concentrate the forming in the center of the die, a wide margin (¾ inch or more) around the forming part is suggested.

Annealed metal from 28 to 36 gauge will permit the cleanest impression when embossing. Metal thicker than 28 gauge does not emboss as well because its thickness prevents the metal from bending into small enough radii.

It will take some experimentation to determine how to get the best result from your embossing die, because each one is different. Thin (⅛ or ¹⁄₁₆ inch) urethane pads of either 80 or 95 durometer should be used. They can be layered together (for instance to make a ³⁄₁₆ inch pad), and can be trimmed with scissors to the size of the die to increase their effectiveness. As with matrix dies, the sandwich can go into the press either way, as long as the metal is between the die and the urethane. When I am using embossing dies I often use blocks of acrylic as spacers to take up some some of the extra space between the platens. Hydraulic jacks seem to be less effective when the ram is fully extended, and embossing generally requires close to the maximum pressure that is available from a 20 ton jack.

Conclusion

This information should give you a basic understanding of nonconforming dies. There are many exciting ways to combine and layer these processes. Here are a few possibilities:

- Folded and fold-formed metal can be used in matrix dies. For instance; fold, anneal, unfold, anneal and then press. Folds can be left "soft," or made sharp by hammering. Or, specific parts could be hardened by hammering before forming in a matrix die.

- Etched or roll printed metal, married metal, mokumé, solder-inlayed metal, and other previously "processed" metal can be formed in matrix dies.

- Embossed metal can be formed in a matrix die. The unannealed embossed metal should be placed so that the urethane presses the back of the embossed sheet into the die.

- A matrix die could be made with a carved flange, so that it is also an embossing die. The flange could be embossed at the same time that the silhouette is being formed. (It might be advisable to do the first pressing with a thin pad.)

- An embossing die could be backed up with a punch. For instance, an embossing die made of wire soldered to a flat sheet could be formed and then used in combination with a dapping punch or curved sheet.

I hope that once you become familiar with the basics that you will invent your own particular ways to use dies. Die forming can then become part of the process of metal-smithing rather than an end in itself.

Trouble Shooting

If the metal tears...
Make sure the metal is annealed. Use a heavier gauge metal. Use less pressure and anneal sooner and more frequently. Check the die for sharp edges or burrs. Concave curves may have to be modified.

If the form is lopsided...
Be certain that the plates of the press are level and parallel, that the form is centered in the die, and the die is placed in the center of the press

If the jack fails to provide power after a certain point...
Avoid working in the top inch or two of the press. Use acrylic spacer blocks to keep the ram operating at mid-range where it is most efficient. Check the "O" ring seals in the jack, and be sure it is filled with the correct amount of oil.

If the form is difficult to press evenly...
Sometimes forms contain broad and narrow areas. In a star, for instance, the center forms well but the points remain flat. Use a thin 95 pad as a first step, before going on with thicker pads. It is the nature of the process, however, that urethane will always form deeper into wider places and shallow in the narrow places, with both embossing and matrix dies.

After heavy use the die edges of the matrix die are starting to wear down...
This is normal. Either face the die with a brass or steel plate, or make new plastic die.

SUPPLIERS

Press, Accessories, Hydraulics, Urethane, Acrylic:

Bonny Doon Engineering
250 Tassett Court
Santa Cruz, CA 95060
(408) 423-1023
(800) 995-9962

Distributed by:

Frei and Borel
125 Second Street
Oakland, CA 94604
(415) 832-0355
(800) 772-3456
Fax (415) 834-6217

Hydraulics:

Enerpac
1300 West Silver Spring Drive
Butler, WI 53007
(414) 781-6600 (for a local distributor)

Further die forming information:

Susan Kingsley
Hydraulic Die Forming for Artists and Metalsmiths
20-Ton Press
P.O. Box 222492
Carmel, CA 93922

Susan Kingsley is an artist, metalsmith and teacher. She is the author of *Hydralic Die Forming for Artists and Metalsmiths*, and lives in Carmel, California.

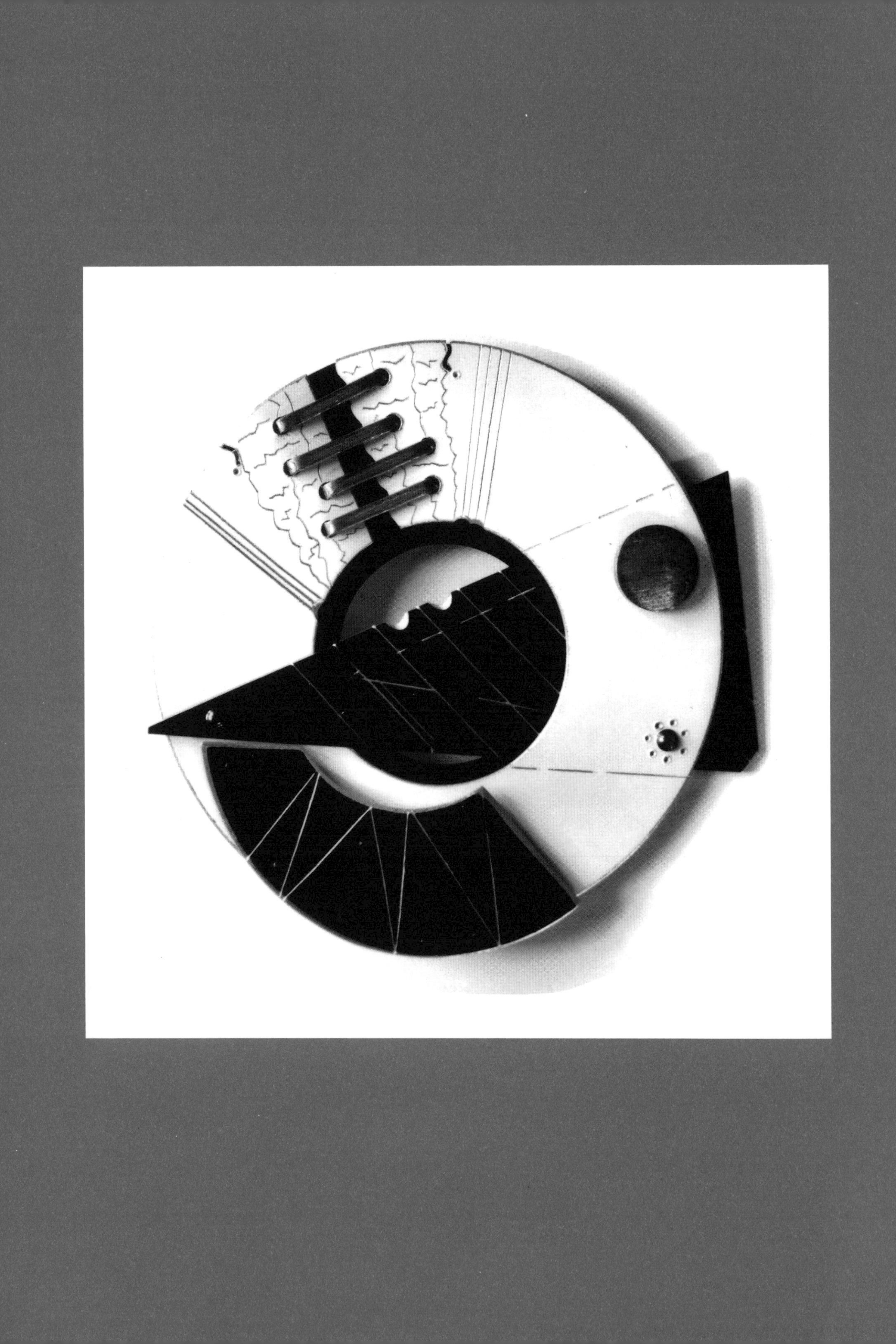

David LaPlantz

Cold Connections

Welcome to a new, exciting and perhaps unlimited way to speed up the construction of your art! Cold connections are mechanical or chemical devices that allow you to join elements without the heat of soldering. In addition to the possibilities of incorporating materials that cannot be soldered such as plastic or paper, cold connections make it possible to assemble prepolished and work hardened units. The best cold connections are not seen simply as necessary compromises to make assembly possible or easier but as a vital element in the design and logic of making art. For me they are a primary creative avenue down which I travel as I make my art.

Perhaps the best way to begin is to discover the many obvious and not-so-obvious cold connections around us. Be prepared to liberate your mind and your materials as we explore the fascinating world of cold connections! I'll bet that after you start using these fastening schemes you'll notice a multitude of connections in everyday life that have applications to your jewelrymaking. Be careful to look closely. It's funny and sad how we overlook basic things in our lives, failing to see relationships and their connection to our work. As I type this into my portable computer I recall slipping the disk into its slot with a click – the cold connection. They are absolutely everywhere!

To assist in your examination of everyday cold connections check out an incredible book called *The Handbook of Fastening and Joining Metal Parts* by Vallory H. Laughner and Augustus D. Hargan (McGraw-Hill. 1956, LCCN: 54-8801). If this 620 page book with its amazing illustrations doesn't give you at least a few ideas, nothing will!

With editing assistance from Shereen LaPlantz, to whom I am forever grateful !

Thinking in Tangents

Before we get into the specifics of making these devices a few general observations might be in order. Thinking (or how we were taught to think) unfortunately often involves moving ideas in straight lines. Personally I prefer to think tangentially. A tangent, as you may recall from geometry, is the turning aside from a straight line or digressing. By thinking in a straight line, life and its solutions are generally boring and less creative than they might be. By thinking in tangents you are liberated, permitted to break off suddenly from one line of thought to pursue another direction. Obviously you can and will return to the straight line as often as required.

With all cold connections consider placement, spacing, the number and size of connecting units, their color, shape and height. Ask yourself if washers are needed for functional or aesthetic reasons and whether special tools or sequences of operation will be required.

Types of Cold Connections

Later I'll talk about adhesives but for now we'll focus our attention on mechanical connections. These fall into two categories; the type you make in the studio and commercially manufactured units that await your personal interpretation.

In a standard rivet, a bulge sits on top of the sheet as at the left. In a flush rivet, a counterbore, or flared opening accommodates this bulge. If the rivet is of the same material as the sheeet, this is called a 'disappearing' rivet.

Rivets

The theory behind rivets is simple: a rod, bar or tube passes snugly through the materials being joined and extends a little further on each side. A hammer is used to tap the metal back onto itself in a process called upsetting. It is this bulge that holds the stack together.

Though they are usually round, rivets can be any shape you want as long as you don't mind the time required to file appropriately shaped holes. This is true of tube rivets as well as the conventional wire rivet, though you'll need special tapered punches to facilitate the spreading of the rivet head.

Rivets can be set with a head either above the surface of the metal or flush with it. Standard rivets give an added dimension to a piece that might otherwise be boringly flat, while flush rivets are visual treats because they appear to float on the surface. If different colors of metal are used the effect glows. On the other hand if the rivet stock is the same color as the metal being joined the rivets will totally disappear.

There are literally dozens of variations on this description, not counting the ones you have yet to dream up. For instance wire rivets can be cut with a fine jewelers sawblade and spread like a cotter pin. A double cut in the shape of an X will give a further variation. In either case the tip of the wire is simply spread with a knife blade or small screwdriver.

I usually use a small ball peen hammer for upsetting. Each blow from the hammer drives the metal out in 360 degrees. By starting in the middle of the wire and working outward in concentric circles the metal flows quickly and evenly. An easy and effective method to demonstrate the action of the hammer blows is to roll a rod of plasticine clay about the size of your little finger, making sure that both ends are flat. Hold the rod snugly in your fist as shown on the next page and use the ball peen hammer as described above.

Notice the relationship between the diameter of the rivet stock, the drill bit required to make a hole that fits and the drawplate hole that yields wire of this diameter. There are three parts to this relationship and when one part is established the other two are dictated by that one. For instance if you only had one size drill bit, that would dictate the wire gauge you need for rivet stock. Which, if you need to draw down the wire, corresponds to a specific drawplate hole. For example a number 53 drill bit equals about a 14 gauge wire which in turn equals hole #25 in my Joubert "E" drawplate. It's comforting to see how easy the relationships fall into place.

For delicate rivets, the wire can be split with a fine sawblade, making either one or two cuts. The grip is made by forcing the sides apart with a knifeblade.

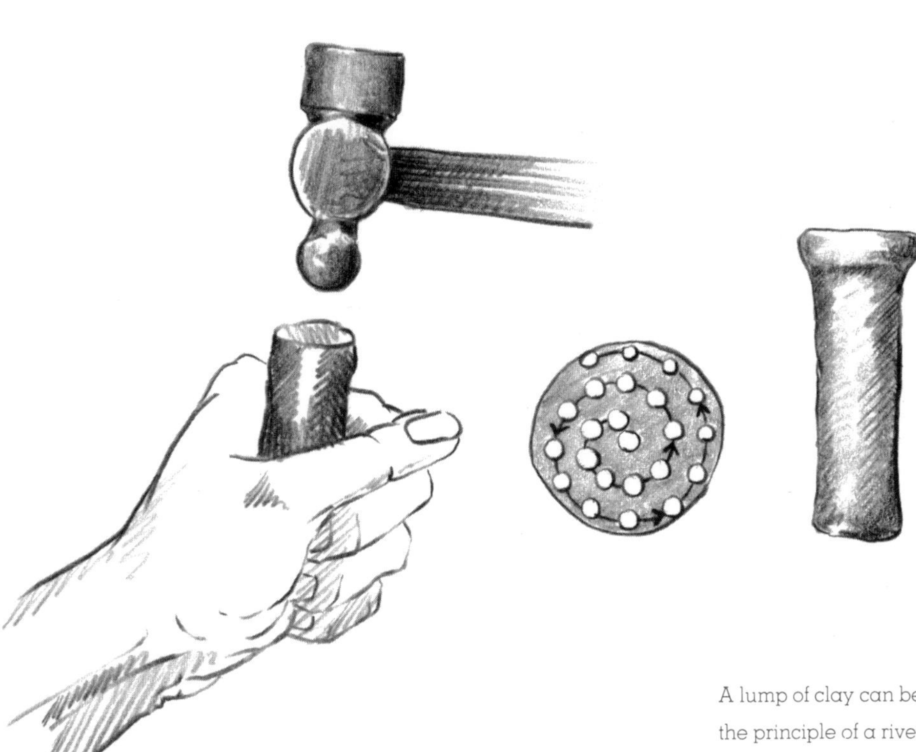

A lump of clay can be used to illustrate the principle of a rivet head. As shown, the hammerblows spiral outward.

Simple Upsetting

1. Grasp the wire in pliers close to the hinge area, allowing about ⅛ inch (2-3 mm) to be exposed.

2. Rest the pliers on the bench pin and strike the wire with a small ball peen hammer, working from the center outward in concentric circles.

3. Keep checking to see that the upsetting is even, both from the top and in profile. Turn the wire completely around to be certain you check it from all angles.

4. Continue upsetting until the head is the desired diameter. Slip it into the hole in your piece for visual reference. Remember that in addition to aesthetic concerns, the rivet head must be large enough to satisfactorily secure the pieces being joined.

5. If many rivets are required, make them all now when you're into the rhythm of the operation. In this way they are more likely to all look the same.

6. Once you have all the rivets made and the first set of holes drilled, slide a rivet into place and set the assembly over a steel block. File the rivet if necessary to achieve the correct length, then upset it with light taps in concentric circles. When the first rivet is set the pieces are prevented from sliding back and forth but they can still rotate. Line the pieces up carefully and drill one more hole, then secure that with a rivet. Now the pieces can no longer move about, so it's safe to drill all the remaining holes and set the rest of the rivets.

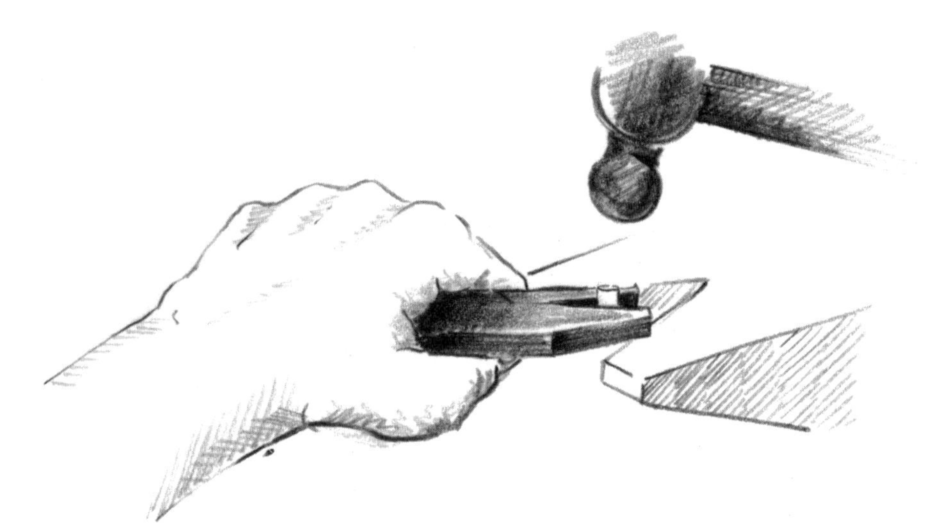

To preform a rivet head, grip the wire in pliers that are then supported against the benchpin.

Some Do's:

1. Know whether your specific rivet material requires annealing before riveting.

2. If you question your design or rivet positioning make a matte board model first. This kind of rehearsal pays big dividends for first time experiences in life or art!

3. Always cut rivet stock longer than needed then adjust length by filing. Cutting new rivets is frustrating and a waste of time and materials.

4. Make a rivet sampler in various materials before assembling your final piece. I'm a believer in spontaneity, but in a controlled spontaneity!

Some Don't's

1. Don't cut rivet stock too short!

2. Don't rivet too close to the edge of your materials. The minimum space is equal to the diameter of the rivet stock. More space is recommended. Remember to take into account filing that may take place after riveting.

3. Don't rivet through layers of material that are too thin. Use thicker metal or add washers to the outside of the pieces. Washers are especially important when joining leather or similarly soft materials that might allow the rivet head to pull through. Of course washers need a hole the same size as the rivet but they do not have to be round or flat. In fact this practical necessity can sometimes provide exactly the decorative element that a piece needs.

In planning the location of rivets,
remember to account for subsequent finishing.

Tube Rivets

1. Drill or punch holes for the tube rivet. A tight fit is important.

2. Anneal the tubing then slide it into place and mark the correct length, allowing half a diameter to project on each side.

3. Saw the desired length, using a tube cutting jig if available.

4. File both ends of the tube smooth and square and slide it into position.

5. Rotate a scribe or similar tapered rod on an angle to flare each end of the tube as shown.

6. Continue the flaring, alternating between the ends of the rivet until it is locked into place. The final spreading of the rivet is done by striking the tube between two dapping punches. Secure one punch vertically in a vise and strike the other with a rawhide mallet.

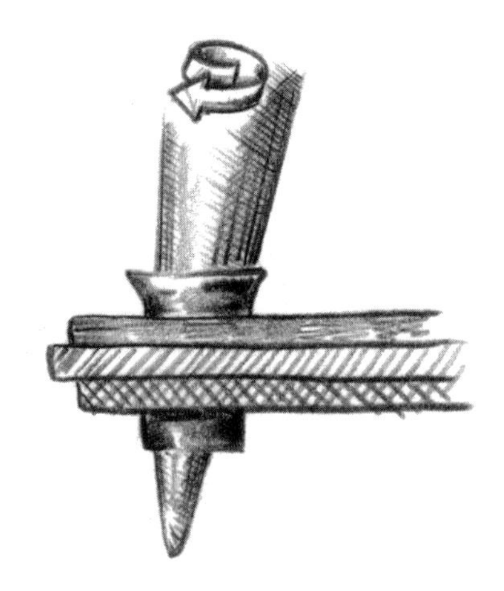

To start forming a tube rivet head, insert a scribe and rotate it gently. Repeat the process from the other side, alternating back and forth as the rim rolls over.

A tube rivet can be further secured by setting it between two dapping punches and striking lightly.

7. A tapered punch with a blunt end is then used to finish the flaring. The other end of the tube is supported on a steel block during this operation and then of course the piece is flipped over and the process is repeated on the other side.

8. Finish the flared surface on each end as needed with files, sandpaper, polishing, engraving, etc.

In an alternate method the end of the tube is slit with a fine sawblade. I recommend making 3 or 4 saw cuts across the tubing to a depth of at least half the amount of the rivet that projects. When the tubing is spread with the tapered tool the saw cuts will facilitate spreading, making the flaring almost instant.

For small tube rivets, make a punch like this to complete the setting. It will leave a tight and neat looking head.

An interesting and especially gentle tube rivet can be made by slicing into the tube with a fine sawblade.

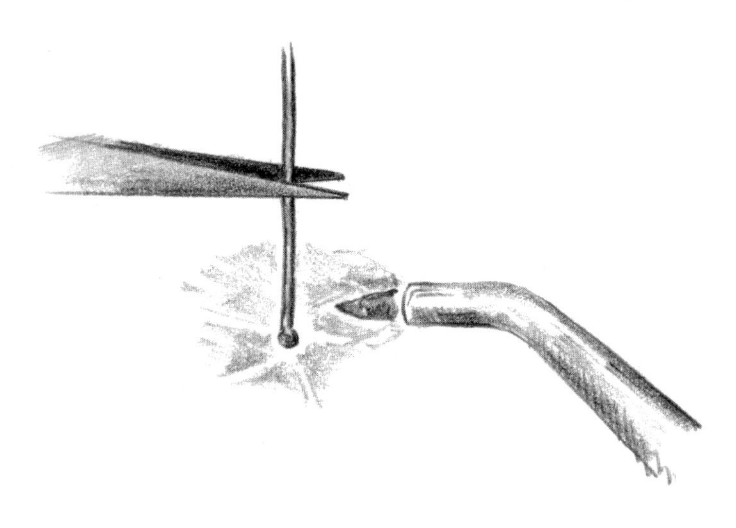

To draw a bead on a wire, hold it vertically in a torch flame. Allow the bead to crawl upward, then remove the flame gradually to insure a smooth surface.

Ball Head Rivet

In this variation a larger than usual head is made on the rivet through the use of heat. Clearly this is done before the rivet is in place. Any jewelry torch can be used to create a small ball on the end of a wire in a process called "drawing a bead." Start with about 5" of wire in the appropriate gauge. Hold the wire vertically 3 or 4" below the plier's jaws and heat the end with a sharp intense flame. As the end of the wire melts, it will draw itself into a sphere and crawl up the wire, getting bigger as it goes. If you try to melt too large a sphere, mean ole Mr. or Ms. Gravity will invite the ball to drop off the end of the wire so I always work over an annealing pan just in case. If the ball is going to drop off it's better that it land in your pan than in your shoe!

The new rivet can be positioned on the piece and riveted as usual with one exception. Having gone to all the work of creating a ball-headed rivet it would be a shame to flatten that head by mashing it on a steel block. Instead, use the endgrain of a piece of hardwood to support the rivet on the underside. An alternative is to use the steel block and hammer with moderate or light blows. The top of the ball will flatten out, leaving the sides nice and round.

To create a nailhead rivet, start by drawing a bead as described above. Insert the wire into the appropriate sized hole in the watchmaker's rivet block or from the front of a drawplate. With this steel support resting on the bench pin, flatten the ball to the desired 'squish' using any flat-faced hammer.

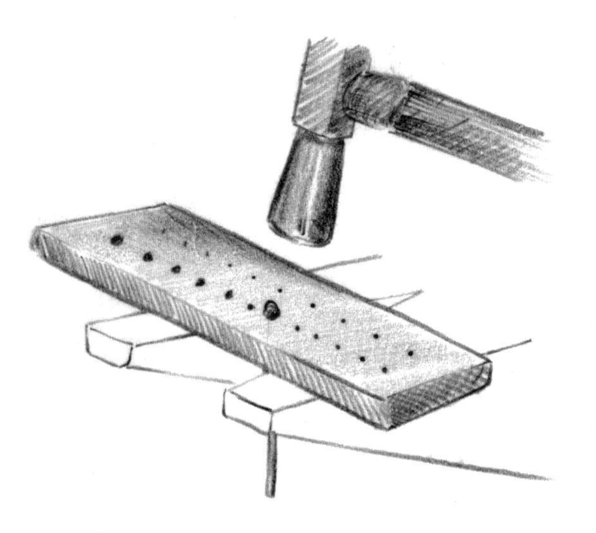

Slide the wire with its bead into a drawplate and planish it to create a nailhead.

Nailheads can be shaped with files, fabricated from sheet and wire, and cast.

Exotic Rivet Heads

Consider sawing flat sheets from 16 gauge or thicker metal sheet, then soldering rivet wires to the back. Suddenly the choice of rivet heads is expanded many fold! Cast forms can also have rivet wires soldered to their backs. Remember that when working with these decorative forms, the support block beneath the rivet should be of wood rather than steel to minimize the damage to the design.

Nails and Commercial Rivets

Brass and iron nails are available in many shapes and sizes. They work well as rivets and are quick to set because one end already has a preformed head, which can be used as is or customized. Commercially made rivets are available in several sizes and materials including copper and anodized aluminum. One end is already flared, usually with a flush head. I am intrigued with the shape of these rivets but they do not lend much creatively to my work. You decide for yourself.

Grommets

Commercially made eyelets or grommets can be used as rivets, assuming they are tall enough to go through all the materials. Your local sewing center should have eyelets in aluminum that are silver colored along with some painted colors such as black, green, white, etc. A simple but effective setting tool can usually be purchased at the same store. One such tool is a pair of pliers with a die built into the jaws. Another is a single male punch that requires a hammer to cinch the grommet. Simple instructions supplied with each rivet setting tool provide gratifying results.

NAILS

A nail can be as strong as a rivet but it must be driven into metal at least ¼ inch thick to provide a solid grip. In my work I like to use HO train scale iron nails but brass nails will work just as well. I have had splendid results using the HO nails to connect 24 gauge aluminum to anodized aluminum rods.

I drill holes the width of the nails through the thin pieces first. Using the drilled sheet as a template I mark the thicker metal and drill pilot holes into the rods. The diameter of these pilot holes needs to be 75–80% the size of the nails to insure a positive grip. They should penetrate only to about two-thirds of the metal's depth. The pilot hole allows the nail to be driven into the metal with ease without it jamming or bending over. Shorten the nail as needed to eliminate the telltale bump that will otherwise appear on the top side of the piece. Cut off nails will need to have new points filed on their ends.

When using small nails, pre-drill a hole that is longer but smaller in diameter than the nail. This will create a friction fit along the shaft of the nail, and insure that you can drive it down tight.

Staple shapes can be used in many creative ways.

STAPLES

The staple is amazing for its simplicity and versatility. Staples, of course, are a one piece construction consisting of a flat top (head) and two legs. Staples are commercially available but in the context of jewelrymaking they are usually made in the studio from either wire or sheet.

If staples are made individually for specific pieces, holes are marked and drilled as each staple is applied. If commercial staples are used or if you have contrived a jig to produce a quantity of identical pieces, make a plastic template to facilitate the placement of holes.

When assembling units, the staple is slid into position and pliers or a bezel pusher are used to bend the legs down. They can be bent either inward or outward. If the materials being joined are not too fragile the staple can be tightened by striking it with a mallet.

If you find your staple connection designs becoming boring, think about overlapping one staple on top of another in an X pattern or perhaps stringing beads on the staple. Staples can be fabricated by soldering wires to sheet metal shapes and cast units, or by epoxying a leg into a hollow form, as shown in the illustrations on this page.

To mark drill holes for unusually shaped staples, trim the ends of the staple legs flat and press them into an ink pad like those used for rubber stamps. Carefully position the staple and lower it against the object, where the ink will leave clear marks to be center-punched and drilled.

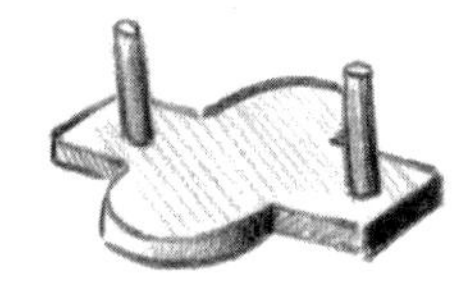

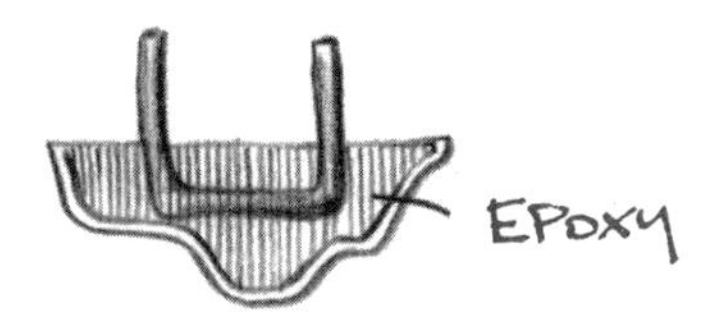

The staple concept can also be applied to fabricated or cast elements.

Coil Connections

The familiar student's notebook presents an excellent example of a spiral connection. Examine one and you'll quickly see a simple theory to add to your cold connection Catalog of Solutions.

To calculate the diameter and length of the spiral, measure the thickness of the material being joined and add about 30%. This is the coil diameter. Obviously this dimension may vary depending on your personal design needs and the degree of movement you want to achieve.

The space between the coil holes and the distance from the edge of the metal should be at least equal to the size of the hole itself. Be sure that the hole is large enough to allow the elements to move freely. A hole that is twice as wide as the coil wire would be typical. Experiment with a cardboard model and you'll soon see the relationships.

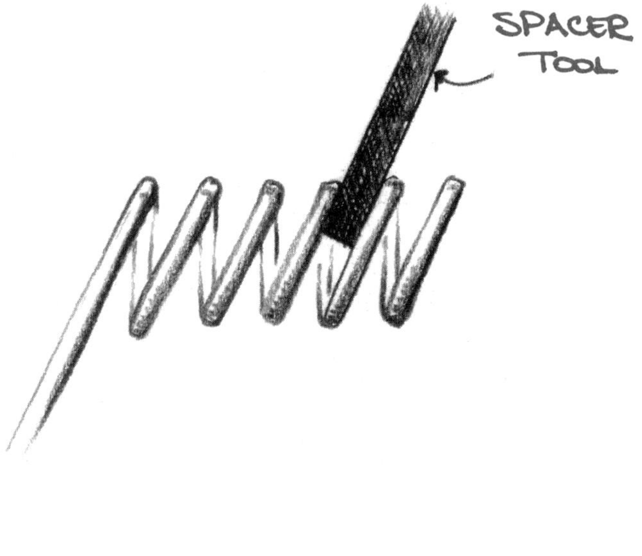

SPACER TOOL

How to Make a Coil Connection

To determine the length of wire for your coil, first decide on the intended diameter. Because the wire needed to make a loop is about three times its diameter ($C=\pi D$), multiply the diameter by 3 to determine the length of wire needed for each loop. Multiply this figure by the number of holes in the pieces being joined and you'll know the length of the wire you should start with. Of course it's wise to add a little more to be on the safe side.

The coil is made by wrapping the wire around a metal mandrel of the correct diameter. Wooden mandrels can compress during winding, making an uneven coil. Grip the mandrel horizontally in a vise so it sticks out about an inch longer than the length of the coil you wish to make. Rather than start from the very end of the wire, allow about an inch to overhang the mandrel when you lay the wire perpendicularly across it. While holding this section, use your other hand to wrap the wire as tightly as you can. Be certain each new turn of the coil is pressed against the last link. With the coil still on the mandrel, grasp each end and gently pull them outward. To guarantee that each wrap is even, make a tool from a strip of wood or stout cardboard. This should be a square about 2" on each side and exactly as thick as the space you want between each turn of the coil. Note that this is also the distance between the holes on the metal plates. Use masking tape if needed, to thicken the tool.

Remove the coil from the mandrel and insert the cardboard tool between the first and second turn of the coil. Rotate the coil, "screwing" it along the tool from one end to the other until the coil spacing is uniform. To assemble the pieces, thread or rotate the coil through the holes, almost as if you were driving a large screw. Bend the wire in a 90° angle at each end of the coil to keep it in place.

Sewing and Stitching

Just as you can sew cloth to cloth with thread, you can also sew metal to metal with wire, thread, or any other kind of fiber. Consider that wire can be drawn to any thickness and comes in many colors and you'll agree that the solutions are limitless. I recommend examining everything you can find that is sewn together. Once you have all this visual information in your gray matter, the solutions will pop out when needed! For more examples look into Oppi Untracht's book, *Jewelry Concepts and Technology* (Doubleday, New York, 1985). He has a great section of stitching examples. Embroidery, basketry and macrame books will also offer interesting ideas.

Tabs and Slots

Remember the tongue-like shapes that hold clothes on paper dolls? Tabs like those and variations on them can offer dozens of possible cold connection options. An examination of toys, appliances and household items will probably reveal many possibilities.

The most familiar use of tabs is to bend them over an edge or through a slot. If the tab has passed through a slot a simple twist will hold the metal pieces together nicely. As mentioned before, a light tap with a mallet can be used to tighten the joint as long as it won't hurt the materials.

Consider making the tabs as decorative as possible, opting for nifty shapes rather than simply staying with the expected tongue shape. Similarly, the relationship between the shape of the slot and its location on the piece can create a sense of drama and personal artistic judgement.

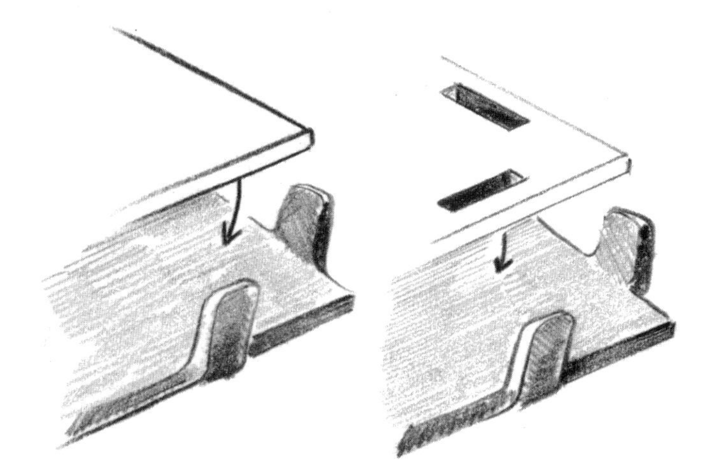

Tabs can be wrapped around the outside of a piece, or can slide into slots.

Adhesives

Though there are contexts in which the use of glue is seen as poor craftsmanship, it would be closedminded to disregard this branch of cold connections out of hand. Sophisticated adhesives are available today that were science fiction less than a decade ago. For a broad-minded person, the addition of adhesives to the jeweler's arsenal is as logical as accepting a new torch or hammer design.

When using any adhesive it is vital that the metal surfaces be chemically clean. I use a Scotch-Brite pad which has the advantage of creating a tooth, or roughened surface, as it scrapes off any oxide or grease film. If the object has a finish you'd rather not damage, clean it with denatured alcohol then rinse the piece and allow it to dry before applying the adhesive.

Follow the manufacturer's instructions meticulously! Especially in the case of high-tech epoxies and adhesives, the success of the joint can be dramatically affected by relatively minor infractions. For comprehensive information on adhesives and suppliers, look into *The Adhesives Red Book* in your local library.

Double Stick Adhesive Tape

Double stick or self-adhesive tapes are fast, fun and easy to use for instant and durable connections. These adhesive tapes come in a variety of forms such as rolls, squares, rectangles and dots in 8, 10, 12 and 14 mm sizes. All adhesive tapes come with a thin protective paper liner on each side. This should of course be left in place until the very moment when the pieces are being assembled.

I recommend using adhesive tapes of different colors and thicknesses to add variety to your laminations. Most adhesive tapes are either black or white but I color the edge with a permanent felt marker to meet my needs.

The Scotch brand 3M #4016 Scotchmount-Natural D.C. 1/16 inch urethane foam tape is superb for holding power as is SE-LIN from Gaylord. The latter is incredibly aggressive, paper thin (36 gauge) double stick and is available from Aardvark Adventures, P.O. Box 2449, Livermore, CA 94551–2449. Write to the Industrial Specialties Division of 3M at Building 220-7E-01, M Center, St. Paul, MN 55144–1000 for information about the Scotch Joining Systems, A-20 Acrylic Adhesive Family Products.

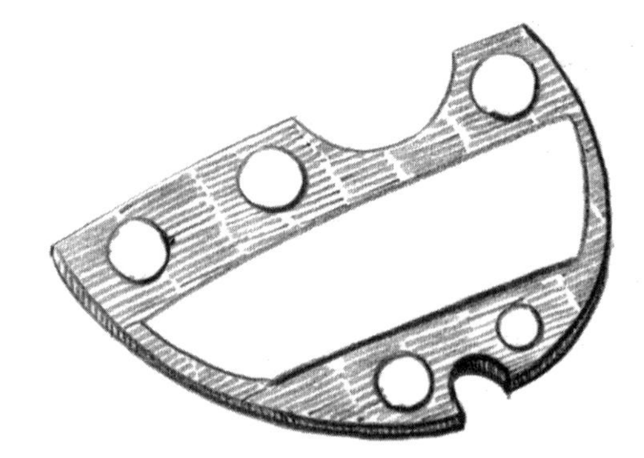

Combinations of dots and rectangles of adhesive are a simple way to secure a complicated shape.

Using Tape

Use a sharp pair of scissors or an X-Acto knife to cut the tape slightly smaller than the object being adhered. If the blade gets gummed up with adhesive, clean it off with Q-tips and acetone. Be careful that you don't accidently cut yourself.

As mentioned, be certain that the metal to be joined is clean and grease-free. Because oxides form quickly, the metal should be cleaned within a few minutes of the construction. I always place the tape on the back side of the top or overlay piece first because it's usually easier to position the element from the top down. I use my knife blade to remove the protective liner and to help in the placing of small pieces. In fact I find that I rarely cut odd shapes of tape, but rather cut a square or rectangle for the body of the piece and rely on precut dots to take care of the odd shaped extensions.

Screws

Small taps and dies are available to the goldsmith who wants to make his own threaded units but I use commercially available brass screws and nuts. These are sold at hobby stores that cater to model railroad enthusiasts and come in a range of sizes and lengths. The same source will also sell socket tools that will make the tightening of these tiny elements relatively easy. It's a worthwhile investment if you plan on using this type of cold connection a lot. If not you'll be able to get by with pliers or tweezers.

To use the screw, drill a hole that is large enough to allow the threaded rod to pass through without having to be forced. This could damage the threads. Make certain that the screw is long enough to allow 4 or 5 threads to stand proud of the surface. If washers are to be used, remember to calculate them into the overall length. Brass washers are usually sold along with the screws but of course you may want to make your own.

Assemble the pieces and tighten the nuts. Excess screw material can be snipped and filed away, which has the added advantage of creating a bur that will lock the nut in place and minimize the chance of its vibrating loose. Of all the cold connections mentioned, this has the unique advantage of ease of removal. For this reason it is recommended when there is the likelihood that the jewelry piece will need to be disassembled in the future.

Conclusion

The cold connection should not be seen as a secondary method for constructing your art. Think of it as the primary creative avenue down which you may travel to partially or completely assemble your art. The time required to construct your pieces with cold connections might allow you to have more free time to enjoy your life. And to make more art! Hallelujah!

David LaPlantz is a designer and jewelrymaker whose work can be found in numerous collections and galleries across the United States. He is the author of *Artists Anodizing Aluminum* and a professor at Humboldt State University in northern California.

Charles Lewton-Brain

An Introduction to Fold Forming

Fold forming is a system of forming techniques that take advantage of the inherent properties of metal. The term fold forming refers to a group of related procedures used primarily on sheet metal. In this article I'll introduce fold forming and describe several specific procedures in a recipe-like manner.

Fold forming is extremely efficient and very rapid, often effecting radical changes in form and surface within a few minutes. The tools used are simple; fingers, hands, hammers, mallets, anvils and a rolling mill. Complex high relief forms can be produced from single sheets often with only one annealing. The techniques may be used with most metals including aluminum, niobium and steel. Metal in the 24–26 gauge range (.4–.5 mm) works well, but fold forming can be done on almost any thickness of metal. I've seen a line fold worked on the inside of a bowl 3mm (⅛ inch) thick. Because no soldering is involved in fold forming, the surfaces and forms produced are useable for jewelry, holloware, enamelling, anodizing and so on. I do most fold forming experiments in thin copper (.016"), sometimes called *roofing copper* or *16 ounce copper*. Precious metals can be used in thinner gauges for equivalent strength. Sterling and gold alloys work well in 24–30 ga (.5–.2 mm).

Fold forming works closer to the limits of ductility and malleability than most standard metalworking techniques. It produces shapes that are reminiscent of forms in nature (ram's horns, ferns, etc) largely because the process exploits the metal's inherent tendencies. This is, in a sense, the way Nature works as well. Certain fold forms reflect natural objects because they develop according to the same rules of physics that direct the growth of organic objects.

Historical Notes

I first presented my ideas about fold forming at the 1985 conference of the Society of North American Goldsmiths in Toronto. Since then the idea has spread to hundreds of people in a number of countries. There has been a steady progression of new forms and methods of working and the potential seems high for further development.

The ground for fold forming was laid in 1979 when I travelled to West Germany to study with Professor Klaus Ullrich at the Fachhochschule für Gestaltung in Pforzheim. He instructed his students to "do things" to the metal; that is, to bash it, cut it, bend it, play with it without an object in mind, and to observe what happened as they played. The metal often produced intriguing and beautiful results. Each student was encouraged to experiment with the most interesting accidents, working to refine both the possibilities and their control of the effect. I was there almost two years, and during that time I experimented with numerous types of surface and forming effects and enjoyed the freedom and rapidity with which ideas could be worked out. At that time and in the next several years when I had returned to Canada, I continued to explore in this direction. The process that most held my attention involved folding, working the metal in some way, annealing and unfolding the sheet. Many of the samples I made were based on a series of experiments I did by folding thin metal sheets, running them through the rolling mill with strong pressure and unfolding them in different ways. In 1984 I went to SUNY New Paltz to undertake graduate study with Kurt Matzdorf, Bob Ebendorf, John Cogswell and Jamie Bennett. It was then that fold forming developed as a system.

Notes

There are some concepts without which fold forming would not have been developed. They have to do with emphasizing understanding and consciousness of process, being aware of the restrictions of preciousness, incorporating serendipity into one's work and examining traditional work methods in the light of process. All this is done in an effort to become more efficient and free to solve technical and aesthetic problems.

Process is what is actually happening when metal is worked. ***Procedure*** is a way of effecting a process; it is a recipe, a technique. There may be dozens of procedures to obtain a similar end effect, but there will be only one process going on. Contrast and comparison of technical procedures and options leads to an increased understanding of Process, the nature of the material used and a freer approach to metalworking. Fold forming derives from a close examination of process in manipulating sheet metal with hammers, clamping tools and a rolling mill.

Metalsmiths are often taught by rote, or what may be thought of as tradition. Tradition is at once a great strength for the goldsmith and a terrible trap. Traditions always exist for a reason and the original reason is always a good one. But conditions change and the original reason might no longer hold good. In order to be able to work in freedom, technical problems and traditional procedures must be reexamined in the light of a knowledge of Process.

The jewelry object is intimate in scale, touched repeatedly during the making process and is often made of materials which are of high intrinsic value or involve a long making time. The high value we place on an object often results in a unwillingness to take risks. I find that copper and other 'throwaway' metals are useful for working out ideas that can then be translated into precious metals.

Serendipity, or the 'lucky accident' also plays a role in metalworking. Because of the preciousness and the intense labor of our work, we generally try to incorporate accidents into a design rather than scrap the piece. By making the accidental moment intentional, by choosing it and altering the piece to match the accident, we bring the work into some kind of balance again. We usually do not stop to realize that we have stumbled upon a design tool. Fold forming has a lot to do with the use of serendipitous moments.

Fold Forming

Fold forms themselves have developed a great deal since 1985. There are now a number of discrete avenues of research available involving folding, working and unfolding. Each avenue provides great numbers of individual 'marks' (specific effects that are potentially useable) or process elements that are distinct enough to be worthy of naming. These are arrived at by following specific procedures – recipes that produce given results.

When metal is stacked in folded, pleated or layered sheets and is worked by rolling or forging, every sheet is similarly thinned or stressed at the same time. This is extremely efficient. I find it interesting that thickness strength can be exchanged for structural strength. A fold form is not necessarily weakened by thinning parts of it.

Because the processes used in fold forming are so close to the limits of the material, the most basic characteristics and tendencies of the metal begin to show themselves in the forms produced. The same laws that dictate form in nature are reflected and reiterated in fold-formed metal, lending the work a type of natural beauty.

The four major families of fold forming are illustrated here. In the following pages several examples of each category will be described. You'll notice that some specific folds have been assigned names. In the early years of fold forming I started the practice of recognizing a major contribution by naming a particular fold after the person who discovered it. This became unwieldy after a while but there continue to be fold forms developed worthy of names.

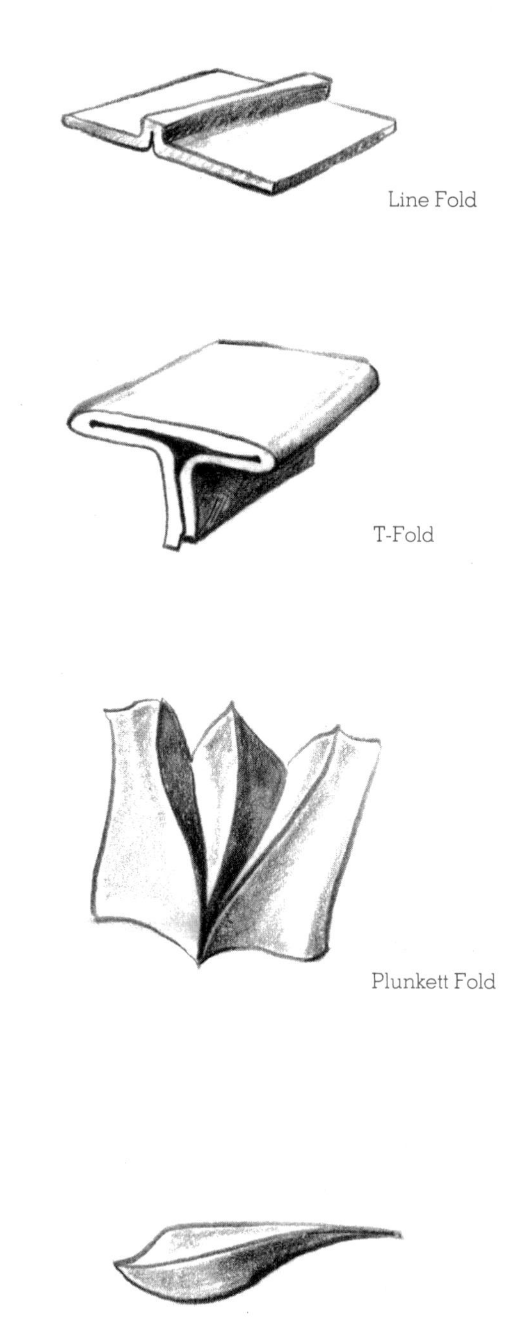

Line Fold

T-Fold

Plunkett Fold

Wire scored fold

Tools and Materials

There are no specialized tools for fold forming. A range of forging and raising hammers in various sizes, an anvil and a solid vise are suggested. A sheet rolling mill is useful for many fold-forms and is essential for a number of them. A raising hammer with a thumb-like face and another with a forging peen are useful. Use goggles when hammering and always check that the hammerhead is not loose. Safety is your business. Think, use your head; live long and prosper.

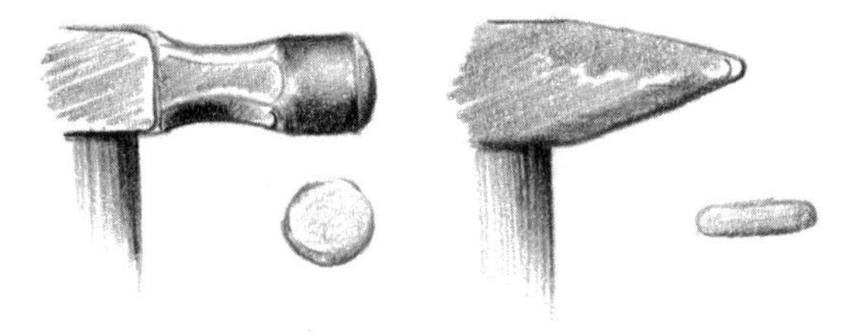

Hammers used in fold forming.
Faces should be rounded and polished.

Line Folds

These folds produce raised lines in flat sheet that resemble chased or even constructed lines. They may run across an entire surface or be restricted to a short strip.

Fold the metal where the line will be placed and mallet the crease flat, as in Figure A. The tighter the fold edge, the sharper the resulting line will be. Anneal and open the form with your fingers. Dry the piece, place it in the rolling mill and take several passes, gradually increasing the pressure to press the line into its parent sheet. Do not, however, crush the line out of existence. This process is called *confirming*. Line folds may be repeated across a sheet and may be made at different heights by using the rolls to compress them differently after opening. When the fold edge is made very tight, the line will look like a square wire soldered on to the sheet. As an alternative to malleting the fold, the entire folded piece can be passed through the mill to tighten the fold edge.

Line folds create a ridge across a sheet, and can cross one another.

FOLD EDGE
LEGS
A
B
C

To make a line fold, bend a piece of metal in half (A), and mallet it tight. After annealing, the sheet is opened (B) and malleted or rolled flat (C).

Centered Line Folds

In this variation, only a section of the fold is malleted closed. When annealed and opened, the result is a line that blends into the sheet on either end. To crimp only a specific area of the fold, pinch the metal between a curving stake surface and a rounded raising hammer face. The short centered line fold that is produced is confirmed; that is planished or rolled into itself as described above.

Forged Line Fold

A piece of metal is folded over and the fold-edges are malleted. A directional forging hammer is then used to forge the fold edge outward, moving from the center out. The piece is annealed and opened. The proportion of leg-to-fold edge dictates the outcome of the fold and is an important decision point in determining a fold's eventual look. The shorter the leg the greater the curvature.

A Rueger Fold
Photo by Charles Lewton-Brain

Rueger Fold

This is an extended forged line fold with a varying proportion of leg-to-fold edge. Make a long line fold and mallet it tight. Cut the metal on a curving line that angles away from the fold. This varying leg length creates a curving helix that results when the fold is hammered. The shorter the legs, the greater the curvature.

Hammering starts at the center and moves outward using a slightly rounded forging peen, staying just inside the fold edge so as not to crush it. Hammer blows should not extend more than one half of the distance from fold to the open side of the fold. The blows are at right angles to the fold. As the forged fold curves, the hammer changes position to maintain a 90° position to the fold edge. I usually use three courses of hammering with annealings between. The fold will curl around during hammering.

After all the hammering is done, anneal the fold and open it carefully without kinking it. Start with a knife blade but continue opening using only your hands. Anneal as necessary for smooth opening. Note that more than two layers can be worked at once.

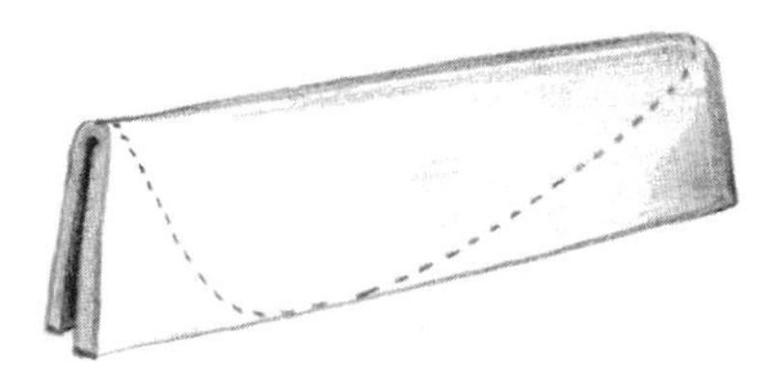

HAMMER BLOWS

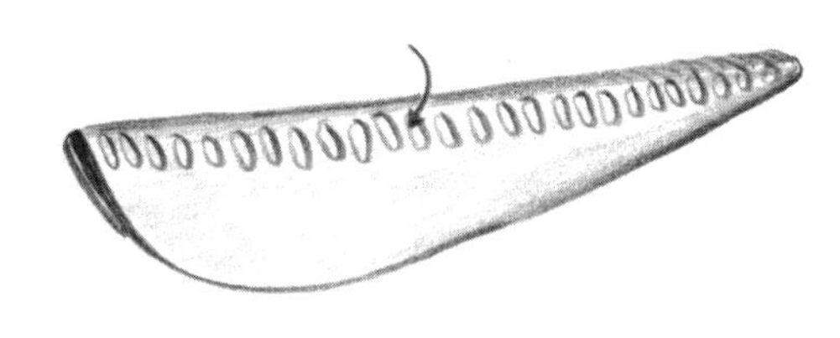

The Rueger Fold is a product of a tapering fold and a planished edge.

HAMMER BLOWS

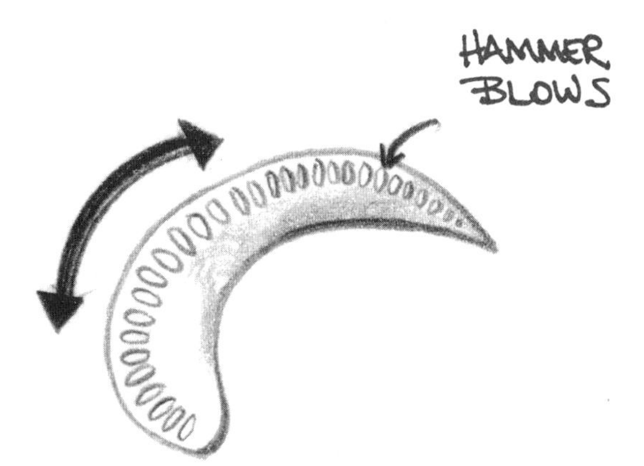

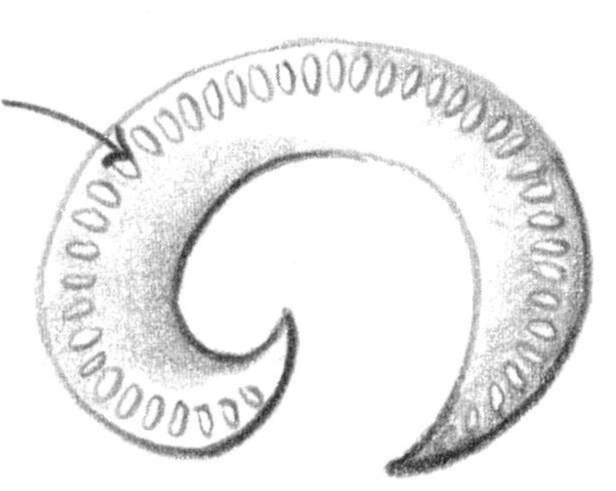

Experiments will illustrate the various effects available with different forging methods.

T-Folds

T-folds represent an enormous number of starting points for an investigation of form. The primary advantage of a T-fold is that two fold edges are formed at once. A basic T-fold is made by placing a doubled strip of metal into a vise so that a "loop" extends above the vise when it is closed. The metal between the jaws becomes the "legs." The ends of the loop are malleted down first (confirming the ends) which leaves the middle of the loop still raised in the air forming a "pillow." This also sets the size and position of the "table" when it is malleted down onto the top of the vise.

Basic T-Fold

A loop is made in the vise, the ends of the loop are confirmed by malleting the projecting section down onto the vise to form a pillow, then the rest of the table is malleted flat. If the fold is opened without annealing, the table stays flat and rigid; after annealing the table becomes concave. Use fairly long legs on a T-fold to ensure sufficient leverage when opening it.

Forged T-Folds

If the entire table area is forged while the legs are pinned in the vise, the table and table returns are thinned simultaneously. When annealed and opened, the table makes a concave curve and the table returns arch over. This points out an important control factor; to increase curvature, the metal sheet should be struck hard and thinned prior to opening. Again, do not be shy about thinning metal in fold forming.

TABLE

TABLE RETURN

LEGS

Left: The steps in making a standard T-Fold.

A forged T-Fold makes a concave curve when opened.

T-folds produce rapid dimensional changes when worked and unfolded. They may be made in various cross sections and with several fold edges. If using more than two fold edges it may be useful to pleat the metal to obtain multiple fold edges. T-folds may be placed in the vise at angles to obtain wedge T-folds and variations on them. The table shape and form can be controlled by using "leg" or "pillow" inserts as illustrated below. Changing the profile of the vise jaws or even changing the position of the fold in the vise while confirming the pillow to obtain a table with curving sides will also produce interesting results. These are the starting points for a number of rolled folds. These folds, like many others, can be worked into objects such as bowls by collapsing the entire bowl, clamping it in the vise to make the T-fold and then opening it. They do not have to be run across a whole sheet but like line folds can be centered by leaving the loop loose on the ends of the fold. A very long T-fold strip can be passed through the vise and shaped up in sections. The vise jaws can be replaced by two pieces of angle iron if the screw of the vise becomes a limiting factor.

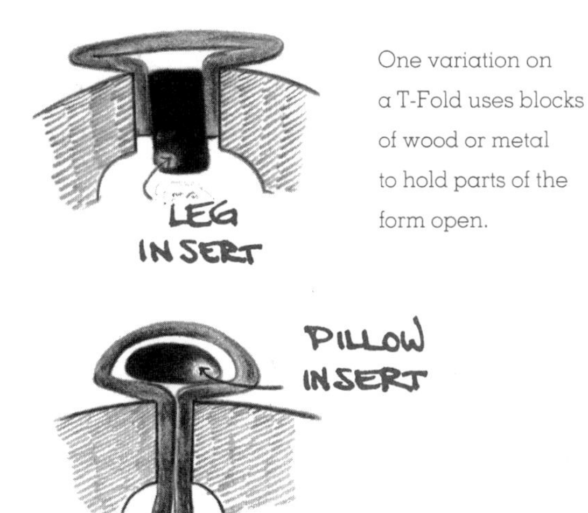

One variation on a T-Fold uses blocks of wood or metal to hold parts of the form open.

T-Folds: Legs Pinned and Free

Figure A (below) illustrates two T-folds made from the same size piece of metal with loops and tables of equal size. In A, the fold edges have been directionally forged with the legs pinned tightly in a vise. Figure B shows an identical T-fold blank where the legs have been forged free, perhaps holding the T-fold against the side of an anvil for the forging. The possibility of having parts pinned versus free offers another important option in fold forming.

T-fold with legs pinned and legs free. Even a small distinction like this can yield considerably differing results.

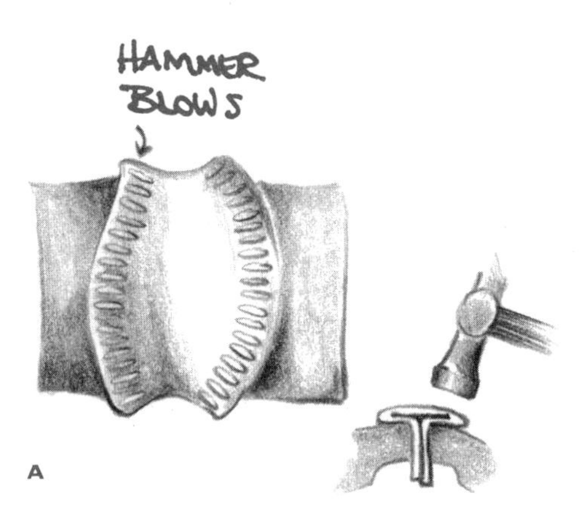

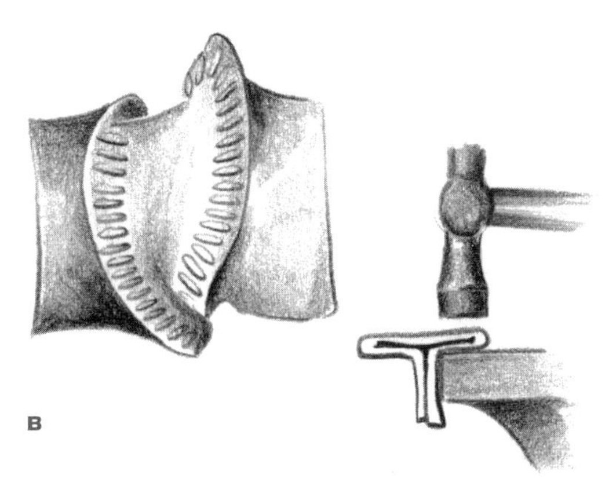

Wedge T-Fold

The loop for the T-fold is placed into the vise at an angle so that when it is malleted into place a triangular table is formed. The fold can be forged and worked in various ways before unfolding. Wedge T-folds can also be placed on both sides of a loop with their small ends meeting in the middle.

Chased T-Fold

Wedge T-folds in particular make good starting points for 'chasing on air.' The idea is to work the loop pinned in the vise with hammers and chasing tools, using the resistance and work hardness of the metal itself as a pitch substitute. As in traditional chasing, broad shapes are worked out first and finer details are worked later in the process. The piece can be removed, annealed, opened, worked out again, annealed and replaced into the vise. Even figurative work can be done with this technique and because no pitch is used the procedure is fast and clean.

Rolled Folds

These folds use the unique properties of the rolling mill to provide equal, directional pressure. I've obtained interesting results with pleated folds and a series of flattened T-folds. These folds, like the "Rueger fold", depend on the stressing of a thick area against sections that have no pressure on them.
This is achieved when one side of a sample is thicker than the other (eg. folded) and it is rolled or forged so the thick side travels further than the thin. Multiple layers made by folding make up thickness that is stressed against the legs of the fold to produce curvature.
In this way a bracelet can be rolled all the way around to form a circle. All rolled folds can be forged, rolled, or worked in a combination of both techniques. Forging followed by rolling leaves hammer texture inside the folds with the outside smooth. Cloth or mesh can be inserted into a fold to make an impression there during working.

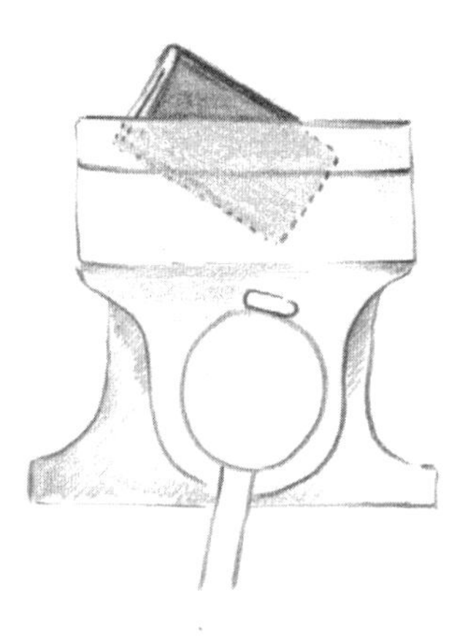

The Wedge T-Fold.

Heistad Cup

Fold a square in half diagonally, mallet it and fold it in half again as shown in the illustrations on the right. Roll it to about two and a half times the starting length by placing the closed corner of the fold in the mill first. Anneal the piece and unfold it, using two chain nose pliers if necessary. Once you get it started, continue pulling it open with your fingers.

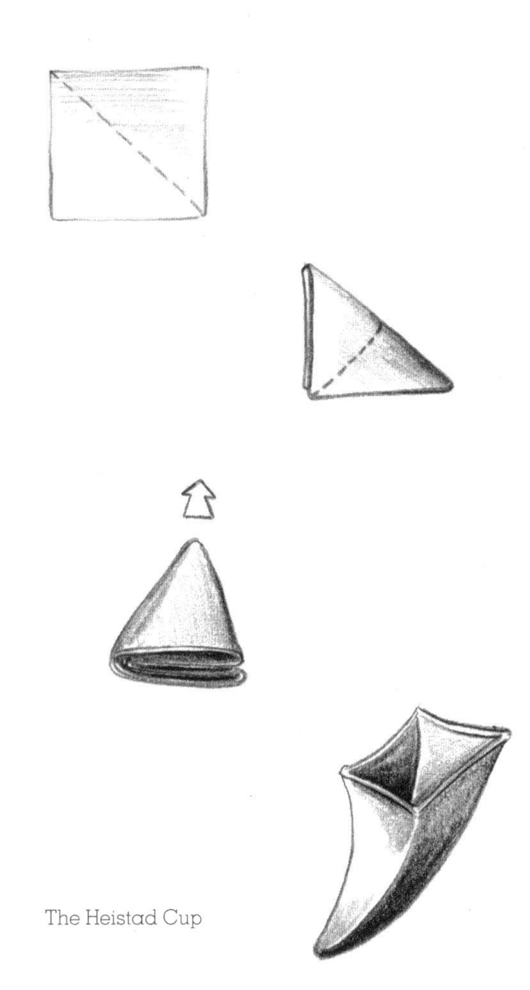

The Heistad Cup

Plunkett Fold

Make a wedge T-fold in the vise, then with one side of the table in the vise, tap the table up slightly with a mallet. Repeat on the other side of the table so that the end cross section resembles a Y. Close the upper arms of this Y and mallet it flat. Roll the fold in a dead pass (no pressure), with the narrow tapered end entering the mill first.

Continue rolling until you approach the plastic limits of the material. If it 'tinkles' you are close to ripping it. If it rips, either you went too far or your original wedge T-fold was uneven which created an overlap. These have a nasty habit of ripping. When done, anneal the form and open it up.

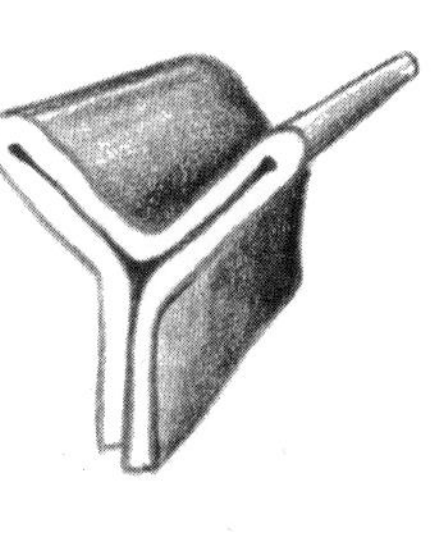

The Plunkett Fold starts with a wedge fold that is closed up on itself. Several can be combined for an interesting effect.

Good Fold

This fold is the same as a Plunkett fold through step 2. After the sides of the table are folded up and prior to rolling, the legs are cut on an angle that tapers toward the narrow end of the pleated metal. When rolled this produces greater curvature, so much that another name for this fold is "the claw."

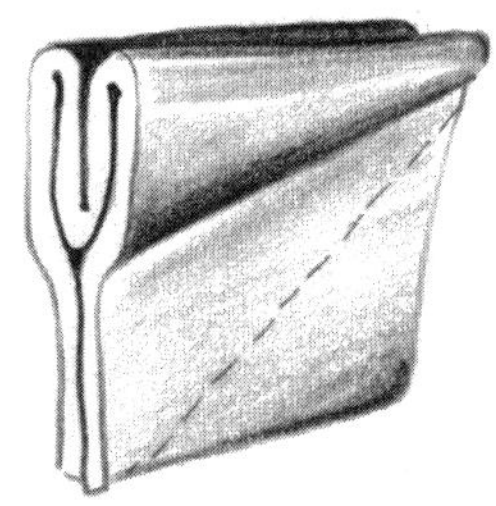

Ward Fold

To make this fold, which is related to the wedge T-fold and Plunkett fold, make a wedge in a long rectangular piece of metal. T-fold this at an extreme angle in the vise, fold the sides of the table up and then forge the fold edges, as shown below.

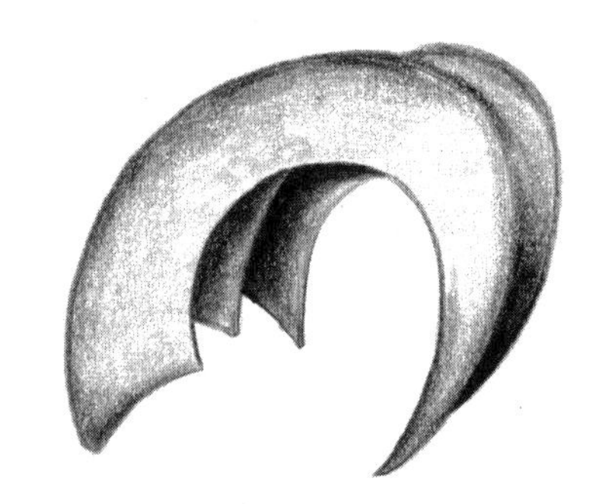

The Good Fold starts like the Plunkett, but is cut to a taper. Forging can be used to accentuate the degree of curvature.

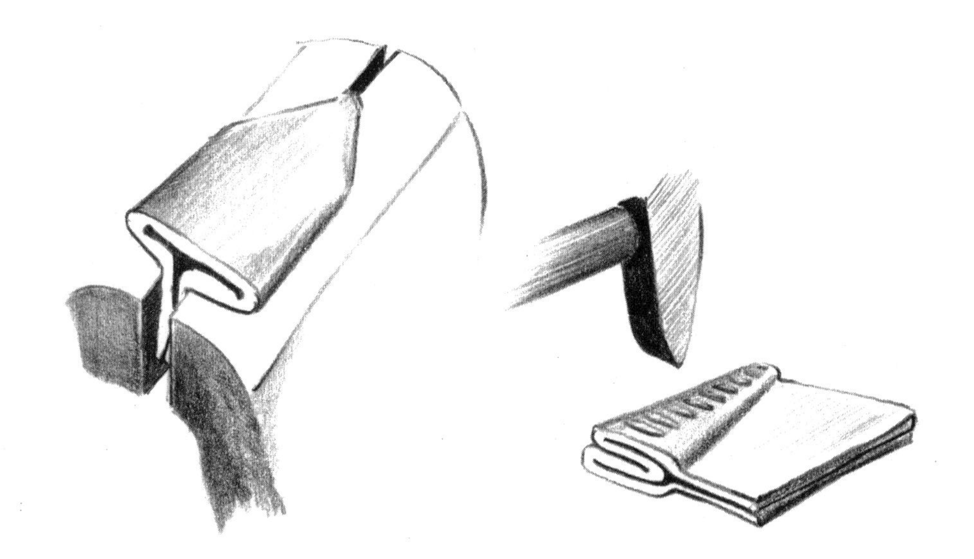

The Ward Fold.

Eckland Fold 1

This is a simple fold in which the edges are forged and then opened.

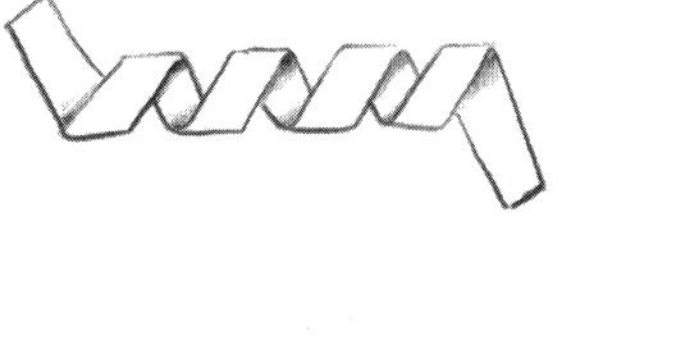

The Eckland Fold is worked on a strip and will change as you vary its length, width and angles.

Circular Rolled Fold

With a wide enough rolling mill and some judicious pulling with a gloved hand while rolling, a straight strip that has been pleated back and forth can be rolled so that after opening it makes a complete circle. It's tricky to get the proportions right and it takes me several tries normally to elicit such a degree of curvature from a strip.

Woven Folds

This recent category of fold forms depends upon interweaving the metal, working it and then opening the fold. Many of these configurations are similar to the gum wrapper chains or scout braiding projects we all learned in grade school. There are vast numbers of unexplored fold forms to be derived from textiles, fabric and basketry techniques, all based on folding and interweaving, working the unit and then unfolding it.

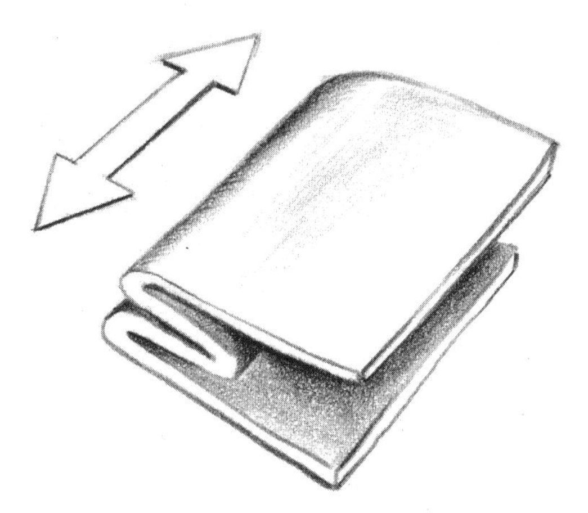

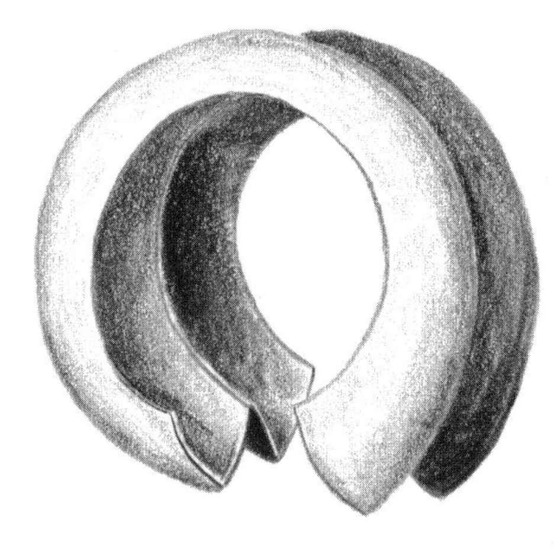

The Circular Rolled Fold. This one takes some practice...

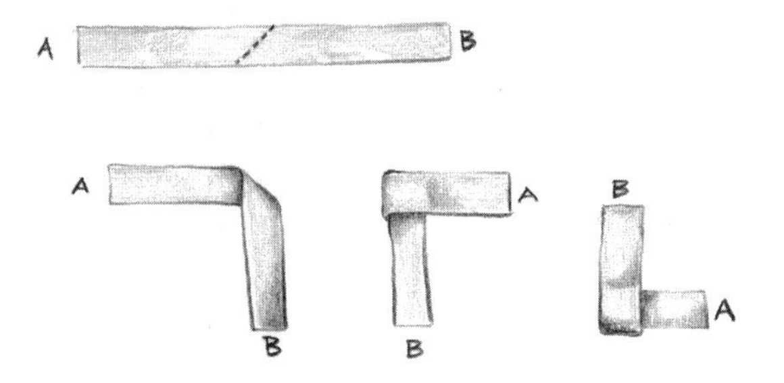

Adams Fold

A long strip of metal is bent diagonally to form a right angle in its center, as illustrated. The two free ends are then bent over each other alternately to make a 'Boy Scout braid.' This can be pulled apart slightly or even forged and then opened up.

Scored Folds

There are a number of folding options that are available only by scoring the metal to bend it, soldering it, and unfolding it. Score until a line shows distinctly on the back side of the metal then bend up with the fingers and solder from behind to strengthen the joint. Bending up a curved scored line will give dimension to the surface relative to the degree to which the curve is folded. By unfolding the piece after soldering and flattening it, you can create curved line folds. Sketches and experiments with scoring and bending can be done with light cardboard such as manila file cover.

Wire can be used to score by rolling it into a sheet. I have used nickel silver, brass and soft steel wire with equal success. Thin round wire, about 22 gauge (.6 mm) works well. Wire scoring is very fast and the lines produced can be sensuous, but they lack the sharpness of conventionally scored lines.

There are of course numerous other ways of scoring and bending including chemical milling (etching), chasing, planishing against an angled edge (Karen Cantine's method) and the use of a computer controlled milling system to carve out the scored lines.

Scoring With Separating Disks

I use straight silicon carbide separating disks to score fold lines, but concave disks are also available. I wouldn't think of doing this without goggles and neither should you. First the bend lines are laid out with a thin marker. I hold the flex shaft handpiece in my fist and brace this against the piece being worked to increase control. Don't press hard, but let the rapidly spinning disk cut its own path through the metal. If the disks wear rapidly, you are pressing too hard. Let the disk drift away from you unless you are left handed, in which case you bring it gently toward you. At the edge of the workpiece the disk will want to run back on the other side. To avoid this, stop before reaching the far edge, lift the disk off, rotate the piece and go in from the edge. Cut until a distinct raised line shows on the back side of the sheet. Make all cuts, then fold the metal gently with the fingers. Some alloys may take several annealings to complete the folding up. If you feel the metal about to break while folding, then flux all cuts and anneal. Do not pickle because this will interfere with later soldering. Place the work in hot running water to remove the glassy flux and continue bending.

Wire can be used with the rolling mill to score a line for bending.

Hydraulic Die Forming

Many fold forms can be opened up in interesting ways by being placed in a die so that the sides of the sheet are clamped. Rubber is then forced into the metal form to unfold the assembly into the cavity of the die. The results look particularly fluid and allow metal flow to become evident. By using a die, a fold form can be unfolded from inside itself in ways not possible by hand.

A re-examination of experiments from 1980 has led to the development of the concept of "forming blocks." These are metal or Plexiglas forms over which metal sheet is pressed using rubber in the hydraulic press. The shapes create areas or lines of work hardness at the edges of the forming block. When the blocks are removed and the unannealed metal is pressed flat using a rubber pad, the work hardened areas act like dies that drive still annealed areas in front of them as they are pressed downwards. Very interesting results that resemble chased surfaces are possible, and can be achieved much faster than with conventional chasing.

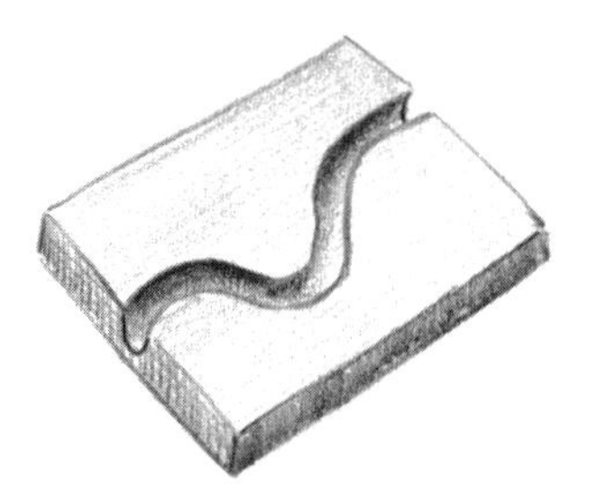

Folds Derived From Paper Models

Because metal flows plastically it will behave differently from paper, but even so, stiff paper can be used to test ideas for fold forms. Working in stiff paper allows rapid investigation of starting folds for metal. It's so easy to work up dozens of ideas in this way that my original policy of naming new folds after the originator has become unwieldy. In a matter of about half an hour you can come up with several folds and starting points for fold-forms. I continue to be amazed at the number of permutations!

To Wrap Up

I hope this brief introduction to fold forming has illustrated some of the exciting shapes that have been added to the metalsmith's vocabulary of forms. I also hope you've discovered the usefulness of fold forming as a device to explore the plasticity of metal and increase your appreciation of the material. By working in copper and saving samples you might also develop the habit of risk taking, even when you transfer the knowledge back to precious metals. I encourage you to schedule an hour a week "playtime" at the bench, working with metal with no end result in mind. I think you'll find that fold forming can be a practical way of initiating this play. I wish you luck and enjoyment!

Charles Lewton-Brain leads an active schedule of workshop teaching, production, writing and teaching at the Alberta College of Art in Canada. With his wife Dee he is a co-director of the Lewton-Brain/Fontans Centre for Jewellery Studies.

Tim McCreight

Toolmaking for Jewelers

A quick look at a jeweler's bench is enough to illustrate the importance of toolmaking for a goldsmith. Punches, chisels, gravers, scrapers and knives are only a few of the common applications of this skill. A goldsmith who knows how to harden and temper steel is able to create a tool perfectly matched to the job in hand. In the case of punches and other tools that make an impression on metal, a large part of the value of a handmade tool lies in its uniqueness. Obviously if your tools don't look like anybody else's your jewelry is more likely to be unique as well. For me it's just as important to add that in making my own tools I am enlarging the process of working at the bench. Because I enjoy making things, any opportunity to expand that process is worth exploring.

This chapter will describe the basic steels of toolmaking and illustrate the process by making a center punch, a stamp and a bench knife. These projects will introduce the skills needed to work steel and will demonstrate several different tempering situations. I recommend that a first-time toolmaker take on the three projects in order.

Historical Background

We know that the Egyptians made knives of meteoric iron that fell to the earth. As you would imagine, metal coming from such a cosmic source was credited with supernatural powers and the simple knives that were made from it were for ritual rather than practical use. By 1400 BC the neighboring Hittites had discovered the basic techniques of smelting, forging and hardening steel. Around 500 BC the Greeks developed a relatively efficient furnace that made steel an economical and dependable commodity. This technology was important in the emergence of the Hellenic culture as a preeminent world power.

In the early days of steel production, in fact through the Middle Ages, billets of steel were flawed by large crystals and impurities. To squeeze these imperfections out, a blacksmith pounded the steel at a high temperature. In order to retain the thickness of a piece of stock the smith would fold the steel over onto itself repeatedly in the same way that a potter wedges clay. This not only created a stronger steel, but as a byproduct, developed the rich linear structure known as pattern welded or Damascus steel. Modern steels are highly refined and do not require this process, but it is occasionally used for tools and knives simply because of the beauty of the patterns.

By definition, steel is an alloy of the elements iron and carbon. If only these two components are involved, the metal is called a *plain carbon steel*. Small variances in the proportions of these two ingredients yield surprisingly different metals. If the carbon is present in amounts of less than .5 percent, the alloy is called *mild steel*. This is the cheapest and most malleable of all steels and is therefore the most common. It is used to make everything from car bodies to safety pins, but is not used for tools because it cannot be hardened.

If the alloy has over 1.5% carbon it is called *cast iron*. This alloy lends itself to casting and is most frequently seen in large steel units that have been cast. Vises, anvils, lawn furniture, and housings for large machines are typical examples of cast iron. Again, this material cannot be hardened and is therefore not used for hardened tools.

Alloys with between ½ and 1½% carbon are known as *tool steels*. They are further classified by a numbering system that uses 4 digits to describe the alloy. The first digit is a code that refers to the principal alloying ingredient. The numeral 1 stands for carbon, 2 for nickel, 3 for nickel-chromium and so on. The remaining digits describe the parts per thousand of this ingredient. A steel called 1075 has 75 parts per thousand (.075%) carbon. A 1100 steel has 1 percent carbon. Either of these would make a durable tool.

Another designation used for steels refers to the medium used to cool or quench the steel in the hardening process. Different steels require different rates of cooling and these are achieved through the use of oil, water or a blast of air as the cooling medium. An "O" steel is quenched in oil, "W" in water and "A" in air. And in case you're not confused yet, there's still another system. Certain alloys have a third or even fourth ingredient to create special characteristics in the steel. These are indicated with a letter code that makes vague reference to their particular asset. For instance the S series is good at withstanding shock and is used to make tools for jack hammers.

Testing Steels

For ease and certainty the best bet is to buy tool steels new from a supplier. While this will cost more than scrap steel it's considerably cheaper than buying readymade tools, so you'll still come out ahead. Of course it's also possible to reuse steel from an old tool. The odds are good that a tool that needed to be hardened (for instance a file, chisel, punch or bit) was probably made of a plain carbon steel in the middle range and has just under 1% carbon.

A time honored test for steels is to hold the sample against a grinding wheel and observe the spark pattern that is created. An experienced tradesman can often make fine judgements based on this test, but for most of us it's simply an aid to determining the right family of steels. The chart below shows some common steel spark patterns. I find it helpful to use a sample of a known steel as a point of reference.

Tools & Equipment

Toolmaking does not require special tools and I generally do all my work at the jeweler's bench with the same files and hammers I use on precious metals. Let it be said here that this is something of a heresy and others would consider this bad practice. I clean my files well after working on steel, I don't try to refine the dust from my sweeps tray and I throw away abrasive paper after it's been used on steel. For hardening and tempering most tools, a jewelry torch is sufficient.

Motor oil is the most common coolant in a metalsmith's studio, but just about any oil will do. Industrially a great deal of attention is given to the viscosity and the temperature of the quenchant, but our less demanding tolerances allow us to fudge on this precision. I generally use motor oil of whatever weight is most easily obtainable, poured into a coffee can with a plastic cover. In a pinch, brine can be used in place of oil as a quenchant.

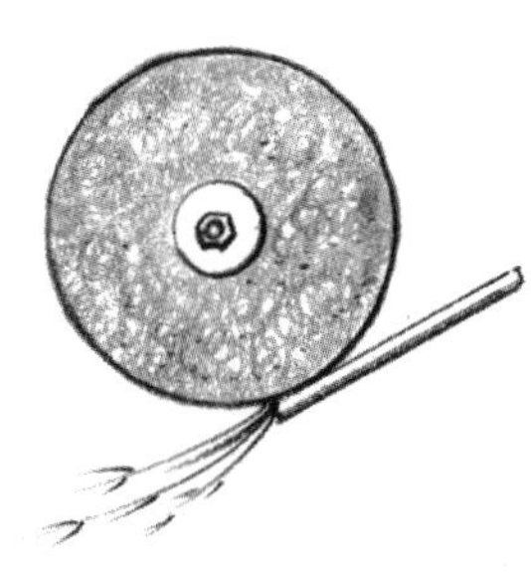

low carbon

medium carbon

high carbon

In a general way the constituents of a steel can be "read" in the spark pattern that results when a sample is held against a grinding wheel.

wrought iron

cast iron

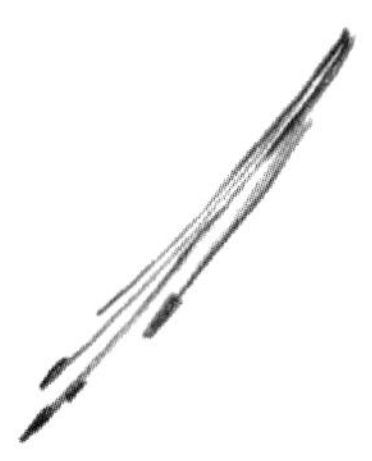

stainless steel

Safety Notes

Hardening steel does not, by itself, entail any special danger. When grinding steel, for instance in the spark test just mentioned, goggles should be worn and loose clothing and long hair should be tied back. In the hardening process a piece of red hot steel will be dipped into a container of oil. If the container is too small, its temperature will rise sufficiently to cause the oil to ignite. This is not as dangerous as it might sound, and usually removing the steel or covering the flame is sufficient to extinguish the fire. The danger is more likely to occur when a startled worker jumps back from the flame or upsets the container of oil. The fumes from hardening are not especially toxic but will accumulate if a lot of work is being done in a confined space. A simple flow of air through the studio is usually sufficient to keep these at tolerable levels. As always, if you are feeling faint or nauseous, stop the procedure and get some fresh air. Some people are more sensitive to gases than others.

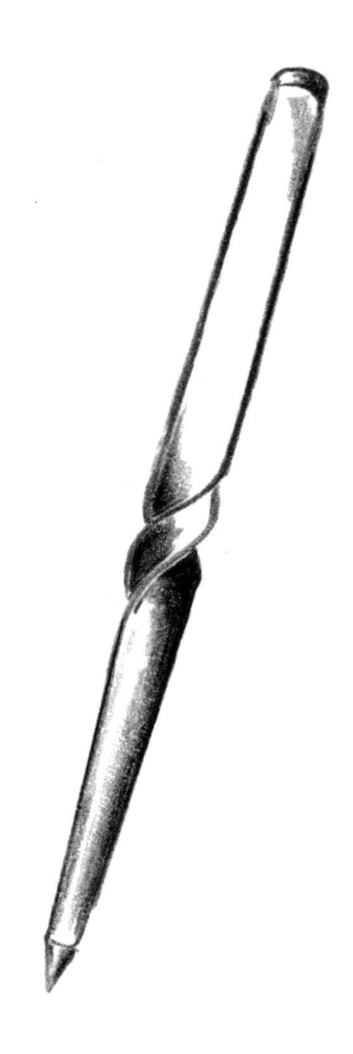

Making a Centerpunch

We'll start with a ¼ inch square rod of plain carbon steel, cut to 4" long. The rod can be cut with a hacksaw, jeweler's saw or a file. To use a file, cut a V notch about a quarter of the way through on each side and bend the bar back and forth a few times.

To make a comfortable grip on the tool we'll start by twisting it in the midsection. Grip one end of the rod in a vise horizontally and secure the other end in locking pliers. Use a torch to heat the zone in the middle of the shaft, bringing it to a clearly visible red. If this zone is short (say an inch) the resulting twist will be similarly confined and therefore tight. If the red zone is larger the twist will have a more leisurely spiral.

With the torch still directed at the rod, rotate the pliers until an attractive twist is formed. If the steel is at the correct heat, this will require almost no effort. As you finish the twist, check to be sure that the planes of the tool are matched up on each side of the twist. Allow the tool to cool down naturally. The heating at this point has nothing to do with hardening or annealing. The steel will be slightly harder than it was before, but this is of little consequence here in the midsection.

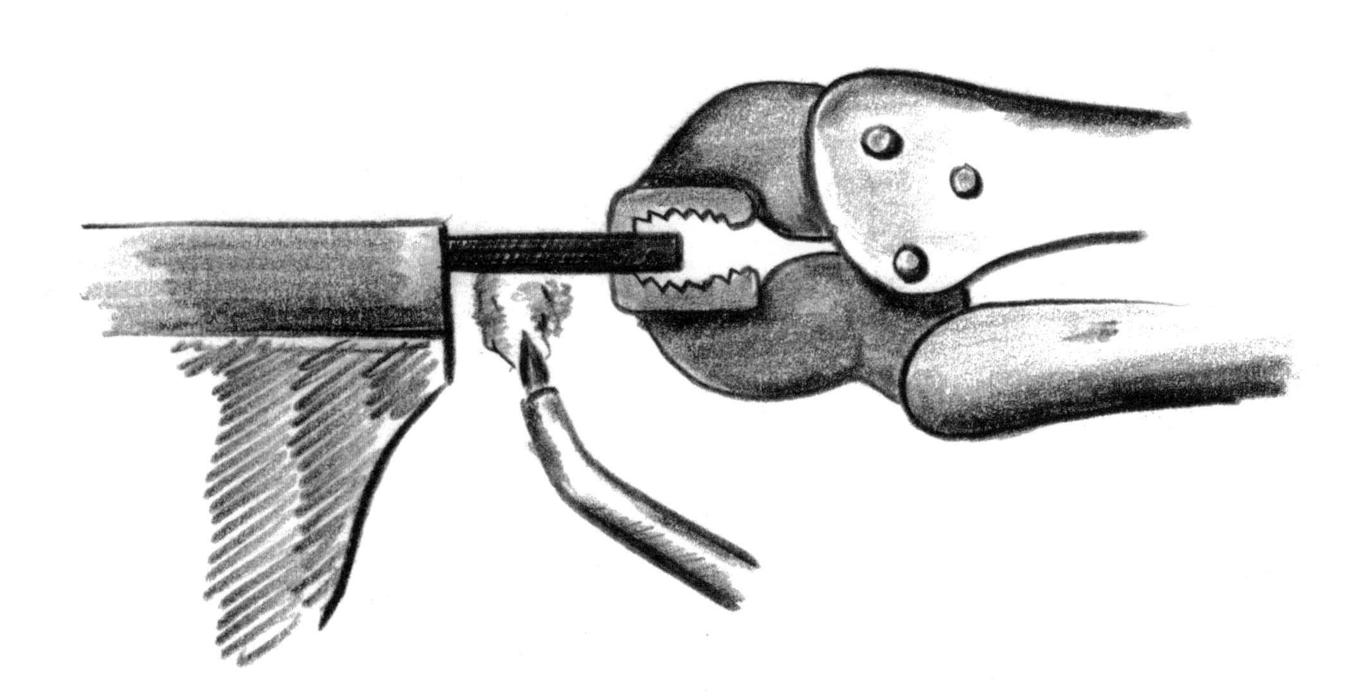

Grip one end of the tool blank in a vise and twist when the steel shows a clear red.

File the top end (the area that the hammer will strike) so the edges are rounded. This achieves two goals. As a tool is used this area will mushroom out as shown in Figure A. The section that curls is workhardened and will eventually shatter, potentially shooting off dangerous chips when the tool is struck. For this reason a mushroom head should always be ground off when it forms. By reducing the area of the tip at the outset you will be delaying the growth of this problem. The other result is that the rounded surface allows a truer contact with most hammer blows. If the top of the tool were perfectly flat, (B) only a perfectly flat blow would make 100% contact.

With the tool gripped in a vise, file a gradual taper on the lower inch and a half of the tool. This will take a while, but the process can be speeded up by using a large coarse file and by standing directly over the vise in a wide stance. In this way the weight of the torso is brought to bear on the file. When the tip is brought almost to a point, move the work to the benchpin and file off each corner of the four-sided taper you have created. This will yield an octagonal section near the tip. Rotate the tool under the file to create a point at the very end as shown in the illustration on the right.

Use sandpaper to remove large gouges in the surface, then grip the tool by its end in locking pliers. With the tool resting on a brick (to reflect back heat) bring the lower 2" of the tool to a bright red-orange heat. This color is easily distinguished after a few tries, but here are a couple tests if you don't trust your eyes to read the color. Set a magnet on a nearby firebrick and touch the red hot tool to it quickly. If the magnet sticks, the tool is not hot enough. If it comes away, the temperature was correct. Another test is to rub the prepared tool with a cake of soap. At the correct temperature the soap will turn black.

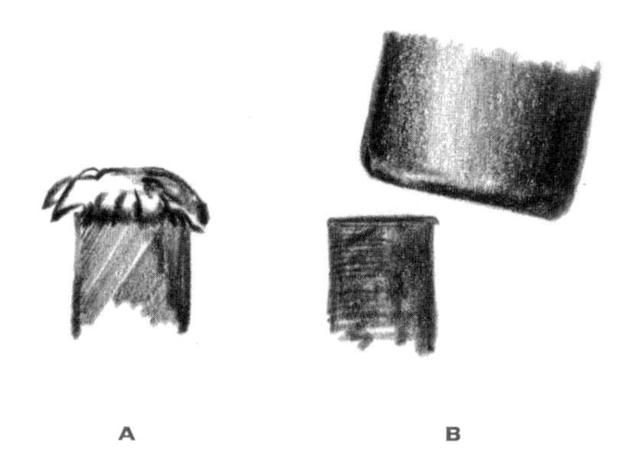

When the top of a punch is heavily worn, it can break apart and throw off small bits of steel. The illustration on the right shows how a flat ended punch presents a corner to the hammer that is coming in at a slight angle. In this case the hammer will push the tool sideways as it strikes downward.

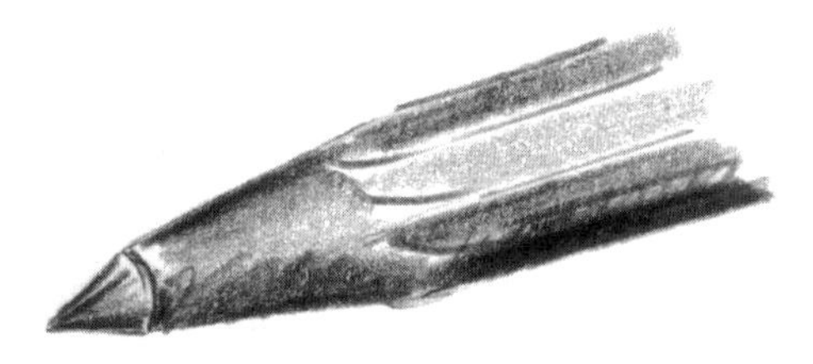

File a general taper on the last inch of the tool, then a sharper taper on the last quarter inch.

Large cementite particles result in a very hard steel, but the small amount of matrix between the spheres leaves the material brittle. In the lower example, the cementite has been partially converted to matrix, creating a tough but more flexible tool.

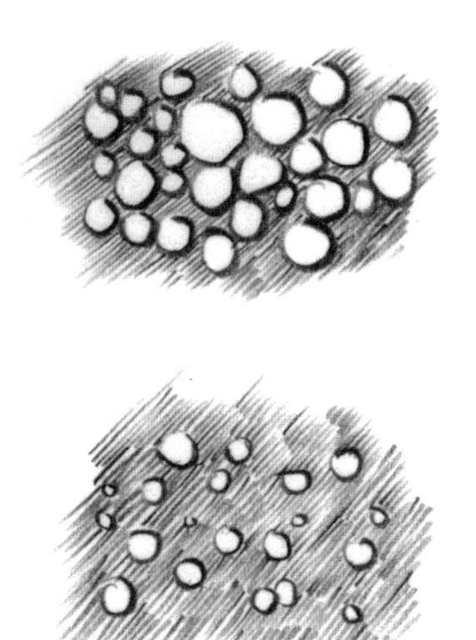

When the tool is at the red-orange heat, (technically called the *critical temperature*) plunge it without delay into a can of oil. As you do this, stir with the tool to insure that it is brought into contact with fresh cool oil. I find it helpful to imagine that the tool is simply a spoon and I'm stirring a pot on the stove.

Leave the tool in the oil for about a minute, or until it is cool enough to be held in the fingers. Wipe the oil on a paper towel and test the tool for hardness. This is done by rubbing a standard file across the tip of the punch. If the hardening was done correctly, the file will make a high-pitched, glassy sound. To distinguish this sound from unhardened steel, slide the file across the opposite (unhardened) end.

At this point the centerpunch is as hard as it can be made, but it is also extremely brittle and would shatter like glass if it was struck. This brittleness is relieved in a process called tempering. Note that you cannot temper without having first gone through the hardening phase. Also, for our purposes you never harden without also tempering.

It might be helpful to imagine hardened steel as a wall made by stacking bricks one on top of another. The bricks are hard but the wall lacks strength because there is nothing holding the bricks together. Tempering could be described as a process in which the outer surface of each brick is converted into mortar. In the case of steel the "brick" is a cementite particle and the "mortar" is called matrix. This magic is achieved by heat; a higher heat will create more matrix and a smaller particle. In practical terms this refers to hardness and flexibility. A steel tempered to a relatively low heat (450° F, yellow in color) will be very hard but will have little flexibility, or "give." A steel tempered at a higher heat (600° F, blue color) will be more flexible but not as hard.

The tempering process is quite easy. If there is a problem it's that the process is so subtle that it's possible to be through it before you know. Start by using sandpaper to remove the black oxide from the tip of the tool. Clean the lower 2" of the tool. The purpose is simply to make it easier to read the colors that will be created in the next step, so it's not necessary to clean away every bit of scale.

With the centerpunch held securely in pliers, direct a soft flame at a point about an inch and a half from the tip. Throughout the tempering, the flame will stay on this point. Nothing will happen for a few seconds, then a blush of color will appear. This bloom will start moving out from the flame in both directions. Ignore the section closer to the top end and direct your attention to the color band that is crawling toward the pointed tip.

You'll see a leading edge that is a pale yellow (straw) color. Behind this is a bronzy-brown, followed by a plum color, followed by a bright blue. For this tool, where hardness is the most important attribute (i.e. we want it to hold its point) the correct temper is indicated by the yellow color. As soon as the yellow gets to the tip, the tool is quenched, either in water or oil. Note that because the tool is tapered there is less metal at the tip and the heat will move faster there. Be prepared to move very quickly as the color band slides down the shaft of the tool. You should be poised over the bath, ready to dip instantly. If you pause to shut off the torch or ask for advice, the moment will be lost. If the tip goes to a brown or plum color, the centerpunch will work, but its tip will round over and need to be reground every couple of months. If the tip is heated past the blue stage you will have effectively annealed the steel and will have to start over by heating it to a bright red-orange again.

Congratulations! You've made a centerpunch that might very well last for several generations. Isn't that amazing! The tool may now be wiped clean and sanded if you want. As a final test, stamp the tool a couple dozen times into a piece of scrap metal, arranging that the last impression is next to the first. Examine them carefully under magnification. If the hardening and tempering were done correctly the marks of the two strikings will be identical. This tool is now harder than a file but can be shaped with sandpaper.

A Border Stamp

In this next project we'll recycle a piece of a file to make a stamp that is useful in creating a border or edge design around a piece. In the process you'll have a chance to practice what you just learned and find out about reusing steel that is hardened from a previous application.

Start with a worn flat file. I manage to create a couple of these each year so they're not hard to come by, but if you need a source you can probably find one at a flea market. We'll want a section about 3 or 4" long and the easiest way to get this is to break off the file. Wear goggles, because bits of steel might fly in unforeseen directions and can be very dangerous. Grip the file vertically in a vise with the correct amount sticking up. Hold a towel or jacket behind the file and give it a sharp blow just above the vise. It will probably snap off cleanly.

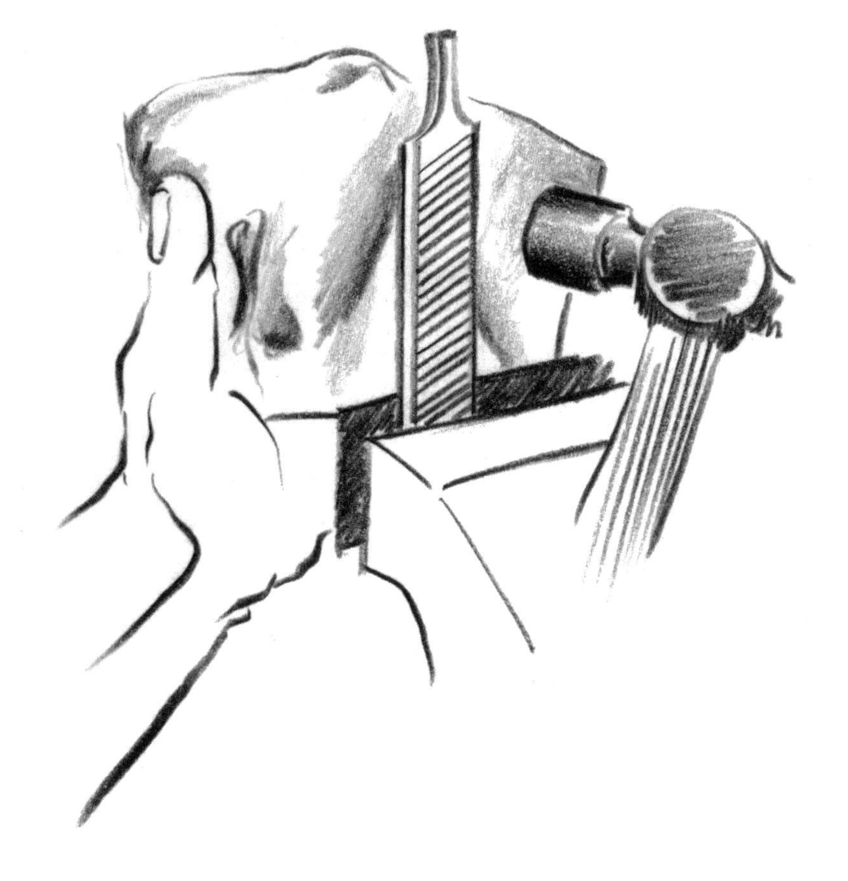

Snap off the file with a sharp blow of a hammer. For safety's sake, catch the pieces in a wad of fabric.

Because the steel is hardened from a previous use it's necessary to anneal it before shaping the tool. Using either a torch or a kiln, heat the steel to a bright red, then cool it very slowly. A traditional method for this is to bury the red hot steel in a deep bucket of ashes or dry sand. If a kiln is used, an alternate method is to simply turn off the kiln and leave the steel in place until the whole unit cools down. The longer this cooling process takes, the more malleable ("softer") the steel will be. A cool down that takes a couple hours is typical. To test the softness of the steel, cut it with a file. If the annealing was complete the file will cut deeply.

Set the tool blank in the vise and file the top and bottom edges flat and square. Remove file marks with a medium grit sandpaper, particularly on the end that will take the pattern. There is no need to smooth the surface of the old file and in fact the teeth will contribute to a positive grip.

Use dividers or a ruler to mark a line down the center of the tool face. With the tool firmly secured in the vise, strike a line of dots with your centerpunch (Got one?) at even intervals along this line as shown below. It's important that these be deep, so you might need to strike a couple of times. If larger holes are needed they can be made with a drill. To check the location and depth of the marks, press the tool lightly into clay or wax.

Use a jeweler's saw and a file to cut away the steel between the dots in a way that creates an appealing pattern. Again, use clay or wax to test the shape of the tool as it progresses. Note that while this doesn't have to be a long process, care in this phase will be rewarded for years to come.

When the tool is properly shaped it is hardened and tempered as described above. If there is a difference between this operation and the treatment on the centerpunch, it is simply that the steel is harder to "read" because the irregularity on the old file has made it difficult to clean the steel. My advice is to pay attention and use a small flame for tempering. When the tool is complete, test it by stamping several times on copper or brass sheet.

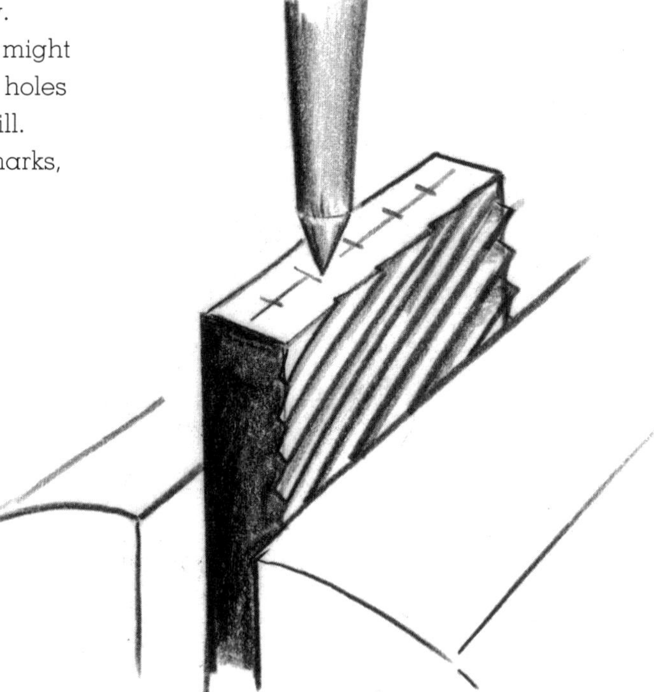

Mark and centerpunch the holes for the border punch.

A Bench Knife

This simple and versatile knife is the precursor of today's utility knife and is characterized by a comfortable handle and a short sharp blade. There are many ways to make a knife like this but for our purposes here I'll try to keep it basic. After a few of these you'll be ready to take on variations of your own contrivance.

Recycled steel can be used for knife blades but in this example I'm going to illustrate with a tool steel sold as *precision ground flat stock.* This is a high quality tool steel available in many thicknesses and widths in 18" lengths. It is usually sold in both air and oil hardening types, of which I recommend the oil hardening as more forgiving. You might be able to buy it locally from an industrial supplier, but a couple sources are listed at the end of this chapter as well. If using a previously hardened tool, anneal it as described earlier.

For a light duty knife like this I would use steel no thicker than 1⁄16 inch. Draw out the knife in its actual size, allowing at least an inch for the tang. From this drawing make a template for the blade and transfer this to the steel. Cut with a jeweler's sawframe using a coarse blade (eg #3) and a little patience. Cut the wooden handle and double check it for a comfortable grip. Use a saw to cut a slice into the handle to accommodate the tang. If a bandsaw is available you might find that it cuts a kerf just wide enough to accept the tang. If it's too narrow, use a piece of sandpaper folded over to enlarge the slot.

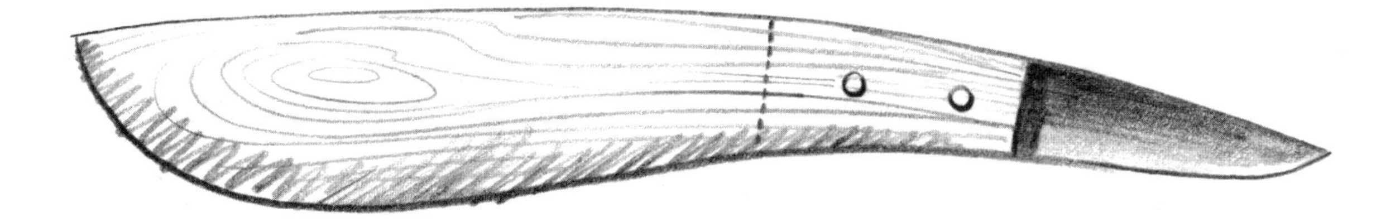

The blade will be held in the handle by two rivets located safely away from any edge and from each other. Mark the locations with a pen and select a drill bit for which you have a wire of matching dimensions. Rivets can be made of copper, brass, nickel silver or steel (use nails). Drill the holes through the wood, taking pains to keep the bit at a right angle to the handle. Slide the steel blank into place and mark the location of the holes with a needle or scribe. Centerpunch the holes (golly, how did I get along without that tool!) and drill through the steel.

To file the shape into the blade, grip it in locking pliers and set those into a vise. To achieve a really sharp edge, file the blade to a slow taper that covers the entire width of the blade. Again, time spent at this step will be rewarded in the hours of use a wellmade tool will provide. When the blade feels about right, switch to sandpapers and remove the marks left by the file. I suggest using a 220 then a 320 silicon carbide paper.

To harden the blade, grip it by the extreme end of the tang in locking pliers. When heated with a torch the thin edge of the blade and the tip will heat up first. To avoid the danger of burning the steel in these areas, direct the torch flame along the back or spine of the blade. When the steel reaches the same red-orange seen in the preceding examples, quench in oil with the same stirring motion mentioned before. When dipping the blade into the oil be careful not to "belly flop" the steel onto its flat surface. This could cause the blade to warp.

Test the hardness of the blade by sliding a file across it and listening for the glassy sound. When you're sure it's hard, sand off the black oxide, remembering that the blade is brittle. If the test reveals that the steel is unhardened try heating it to bright red and quenching again. If it still doesn't harden it might not be a tool steel after all.

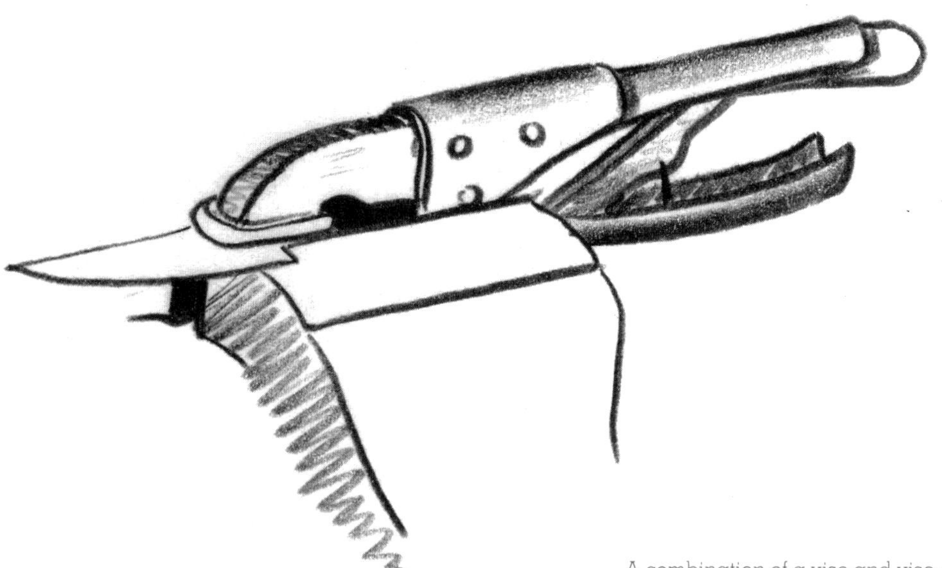

A combination of a vise and vise-grip pliers makes it easy to file the taper of the blade.

Tempering a blade is a little trickier than the punches described earlier. Ideally we want the edge to be the hardest temper (yellow) so it will take and hold a sharp bevel. The back or spine should be flexible (blue) so it can yield as the knife is stressed in use, and the midsection can be plum to combine the best of each of the other two states. The problem lies in the fact that the section that wants the least amount of heat (the edge) is the thinnest and will therefore heat up first.

To compensate for this, set the blade overhanging the edge of a firebrick as shown below. Direct the torch flame at the brick just beneath the blade. This might be enough to properly distribute the heat but if it's not, use tweezers to dip the edge of the blade into water periodically during the tempering when it gets too hot.

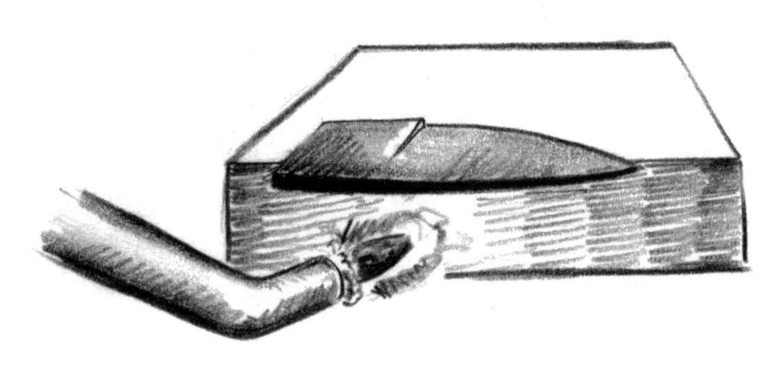

In order to heat the thick spine of the blade without overheating the edge, lay the blade so it overhangs the brick.

When the blade is tempered and cleaned up, the handle can be shaped with rasps and files and sanded smooth. The blade is slid into position and the rivet wires are laid into place. Cut the wire so only a small amount extends on each side and use a small cross peen hammer to shape the rivet on each end, flipping the knife over to work on both sides. Use the flat face of the hammer to smooth and round the rivet heads.

The blade is most easily sharpened after it is securely fastened into the handle. Slide the blade along the surface of a whetstone with some pressure, imagining that you are trying to slice off the top layer of the stone. A light oil should be applied to the stone to float away the tiny chips of steel being created. To hone the edge, strop it on leather that has been rubbed with a polishing compound such as tripoli or white diamond. Jeepers, you made a knife!

SOURCES

Manhattan Supply Company
800 645-7270 or 800 888-7270
800 255-5067 fax

Graingers
800 323-0620 or 800 225-5994
800 722-3291 fax

Tim McCreight is the author of several metalsmithing texts and a professor at the Maine College of Art.

Komelia Hongja Okim

Kum-Boo: 24K Overlay on Silver

Kum-Boo is a Korean applique technique of surface decoration in which pure gold foil is fused onto the surface of finished silver ornaments or objects. The historical background and source of the technique are vague. For thousands of years, Koreans have been using brass, silver and gold ware and utensils for their daily eating and ceremonial purposes. Especially in winter, brass and silver ware are commonly used to keep food at the correct temperature. Since the Korean War ended in 1953, most brass and silver utensils have been replaced by aluminum and stainless steel wares for common use. Beginning in the late 70s however, a revived interest in traditional customs has aroused a new interest in the use of the finer metals, particularly for ceremonial, birthday, and wedding celebrations and for ancestral memorial services.

Among upper class people, silver wares are commonly used to display wealth and to enjoy both their prosperity and status. These wares are also used to test defective, spoiled or poisoned food, and in this regard, they echo a role these metals have played among royal families for centuries. Koreans and some other East Asians believe that the ingestion of pure gold will improve health and well-being. Many herbal medicines are covered with very thin sheets of pure gold. Similarly, acupuncture needles are often made of a high karat gold in the belief that this metal has properties that will enhance the effectiveness of the treatment.

Many Korean silver utensils are decorated with 24 karat gold overlays in the form of letters and patterns that convey wishes for good health, wealth and longevity. In most cases the ornament is set in the interior of a cup or bowl, or within the bowl of a spoon, in order that the food will be in contact with the gold and therefore able to assimilate its positive characteristics before being imbibed.

Though somewhat rare in the United States, and almost unheard of in the commercial industry here, Kum-Boo is a familiar technique in Korea and has seen wide use for the past decade.

OVERVIEW

The term Kum-Boo is derived from KUM (금) which means *gold* and BOO (부) which can be translated as *attached.* Because of the extreme delicacy of the sheet, conventional soldering is not appropriate and would probably be impossible. In Kum-Boo, 24K sheet gold is diffused onto silver alloys through a combination of pressure and relatively low temperatures (500-700°F, 260-370°C). The primary tools used in this technique are a stove, hotplate or torch to create the proper heat, and variously shaped polished steel burnishers that press the overlay against the parent object.

In fabricated pieces, all soldering must be completed before beginning the Kum-Boo process. When the piece is fully assembled (except, of course for elements to be attached with rivets or similar cold connections), the work is cleaned thoroughly in fresh pickle. In order to establish a skin of fine silver, the work is heated until it shows brown spots, then it is again dipped in the acidic pickle. The piece is rinsed in water and lightly scratchbrushed. This cycle is repeated from 3 to 6 times, or until the metal retains its white color even when heated. The work is then dried carefully, with care taken that the delicate skin is not touched.

Overlay pieces are cut from thin 24K gold sheet with scissors or a sharp X-Acto® blade. The gold is laid into place and the work is set onto a hotplate or other heat source and brought up to the correct temperature. The foil overlay is then burnished into the silver, where a diffusion of the molecules of the two metals creates a permanent bond between the two sheets.

EQUIPMENT AND MATERIALS

- Hot plate, camp stove, torch, electric and/or gas stove
- Gum tragacanth powder
- #00 or #1 artist's paint brush
- 24K gold foil .03 - .05 mm
  If Thompson's gold foil is used, double or quadruple layers are used as a single sheet.
- Tracing paper
- Sharp scissors and X-Acto® knife
- Several shapes of curved steel burnishers
- Fine-pointed tweezers
- Cotton gloves
  (not gardening gloves - too bulky)
- Hot pickle (Sparex®)
- Fine-pointed sewing needle
- Baking soda
- Tooth brush (soft bristle)
- Liquid soap mixed with ammonia

Optional:

- Optivisor®
- Tripod
- Hole punches or design punches to cut out repeated shapes
- Steel or brass scratchbrush

Preparation of the 24K Gold Foil

Purchase 24K gold in the thinnest form possible. The desired thickness is .03 to .05 mm, which can be achieved in a rolling mill if necessary. An electric rolling mill is preferred, but a manual mill will work. If the gold threatens to adhere to the rollers, sandwich it between sheets of paper or oxidized copper and pass it through the mill in the conventional way.

After every 2 or 3 passes the gold will need to be annealed. This is most easily done on a hotplate, set on High. If you prefer to use a torch, lay the gold on a sheet of copper, brass or steel to diffuse the heat. Work in subdued lighting and heat until the gold begins to show red, then air cool for a few moments and quench in water.

Note that if Thompson's gold foil is used, limit the number of pieces in an application to 5 or 6. Because the foil is so thin, the prolonged heat needed for more complex applications is liable to diffuse the gold into the silver alloy, where it will almost disappear.

Process

1. Complete the work through all soldering and finishing steps. Patination, stone setting and cold connections can all be achieved after Kum-Boo. Any other procedure should be completed before the overlay process is begun.

2. Draw the pattern onto tracing paper. Fold this to create a sleeve into which the gold foil can be placed. Cut the overlay piece(s) with a knife or sharp scissors. Leave the gold in the paper until ready for use.

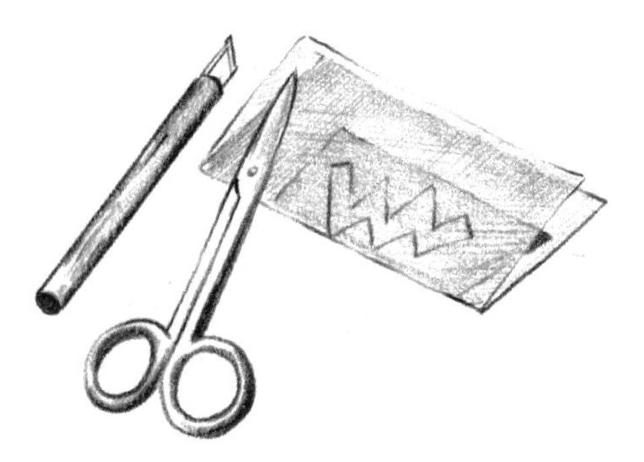

Trap the thin gold in a folded paper and draw the desired shape on the paper. With sharp scissors, cut through the paper and gold simultaneously.

To anneal the gold safely, set it onto a piece of steel plate and use a torch or hotplate.

3. Using a permanent marker, draw the overlay outline onto the work, making the drawing a little larger than the overlay piece. This outline will insure the proper location of the overlay.

4. Place the gold pieces on the work with a dampened brush or burnisher. Simply dip the brush in a dish of water and shake off all excess before using the brush to pick up the gold pattern. If the patterns are large or complicated, use a very thin mixture of gum tragacanth instead of water. Klear Fire, a substitute in the enameling process, should not be used here. I prefer to mix my own solution starting from powder, rather than use the premixed commercial liquid solution.

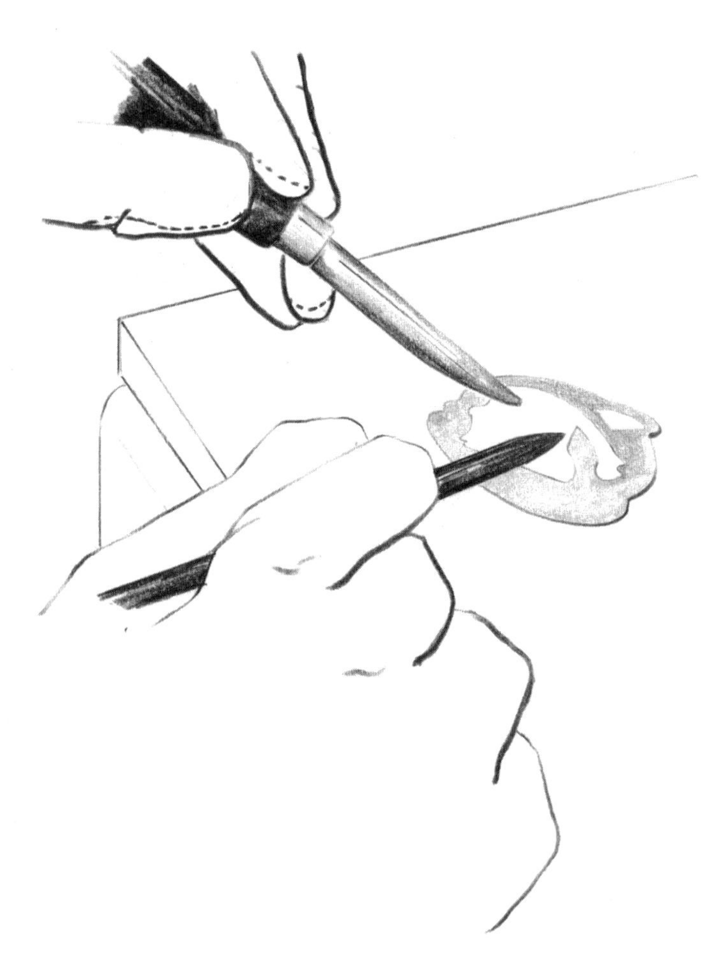

5. Set the work on an unheated hotplate or burner and set the temperature on High. In order to provide a flat surface, you may want to set a piece of steel or a metal screen on the burner and allow this to distribute the heat. As it dries, the water or gum tragacanth will secure the foil to the surface. When the object reaches (500-700°F, 260-370°C), apply light pressure with a burnisher to tack the foil into place.

6. When each section of the overlay is tacked down, lower the heat and burnish the entire piece briskly. It is important that the object be held steady, which might require an additional burnisher or a pair of tweezers. Note that you are working directly on the hot burner; I usually wear gloves to protect my hands from the heat.

7. If the burnisher gets too hot, the gold foil will adhere to it. To prevent this, keep a cup of cold water close at hand and dip the burnisher into it periodically as you work.

8. When the application is finished, examine the edges of the overlay closely to see that they are well attached. A loupe or Optivisor® is useful for this process. If the diffusion was complete, the work is set into hot pickle until the metal is completely white and all burnishing marks are removed. The piece can now be patined or polished, using a rouge cloth or a scratchbrush lubricated with soapy water.

Use one burnisher to hold the gold into position while the other presses the sheet into contact with the silver. This process gets hot, so you might want to wear gloves.

Variations

A. It is possible to overlap layers of foil using the Kum-Boo techniques described above. The gold is so thin it is semi-transparent, which will affect a color change as layers are built up. When working on a flat or nearly flat object, it's most efficient to apply pieces directly onto the object while it's hot. To do this, pick up the prepared piece of foil with a dampened burnisher and carefully set it into position. Keep a slight pressure on the piece until the water stops sizzling, then gently begin to burnish the foil down. When it appears to be tacked into place, increase the pressure and speed of the burnishing action. Remember to quench the burnisher in water when it heats up, and to change burnishers when a new shape is required.

When working on a large piece, a torch might be needed to supplement the hotplate. In complicated forms, gum tragacanth is used to "glue" the gold into position before fusing.

B. Sometimes complicated patterns or unusual contours of the object make positioning of the overlay pieces difficult. In these circumstances, use a thin mixture of water and gum tragacanth to glue the pieces into place. Use a fine paint brush to coat the area that will receive the foil and to pick up the overlay piece. When all the pieces are in position, set the work aside to dry completely.

When dry, set the object onto the hotplate, turn the heat to High, and selectively burnish areas to tack down each pattern piece. Turn the temperature to Medium and burnish each piece fully. Remember to cool the burnisher periodically. When burnishing is complete, allow the piece to cool, then examine each edge for bonding. If the diffusion is complete, set the object into hot pickle.

C. When working on a large object the hot plate might not provide sufficient heat. Diffusion will not occur if the parent metal is not hot enough. In this case the overlay pieces are glued on with the gum tragacanth mixture just described. Use a torch as shown here, often in conjunction with a hot plate. Obviously this process will be easier if you can enlist the aid of a friend.

Repair of Kum-Boo

Poor adhesion is usually indicated by tiny bubbles in the gold. To repair, reheat the area with a soft torch flame, turning the flame aside briefly, then reburnish with considerable pressure. If there are many bubbles present, puncture each one with a fine needle, reheat the area and burnish quickly and vigorously.

Komelia Okim is a professional metalsmith with work and exhibitions throughout the US, in her native Korea and other countries. She was a Fulbright Exchange Scholar to Korea and taught at Hongik University in Seoul in 1982-83. She is a professor at Montgomery College in Maryland.

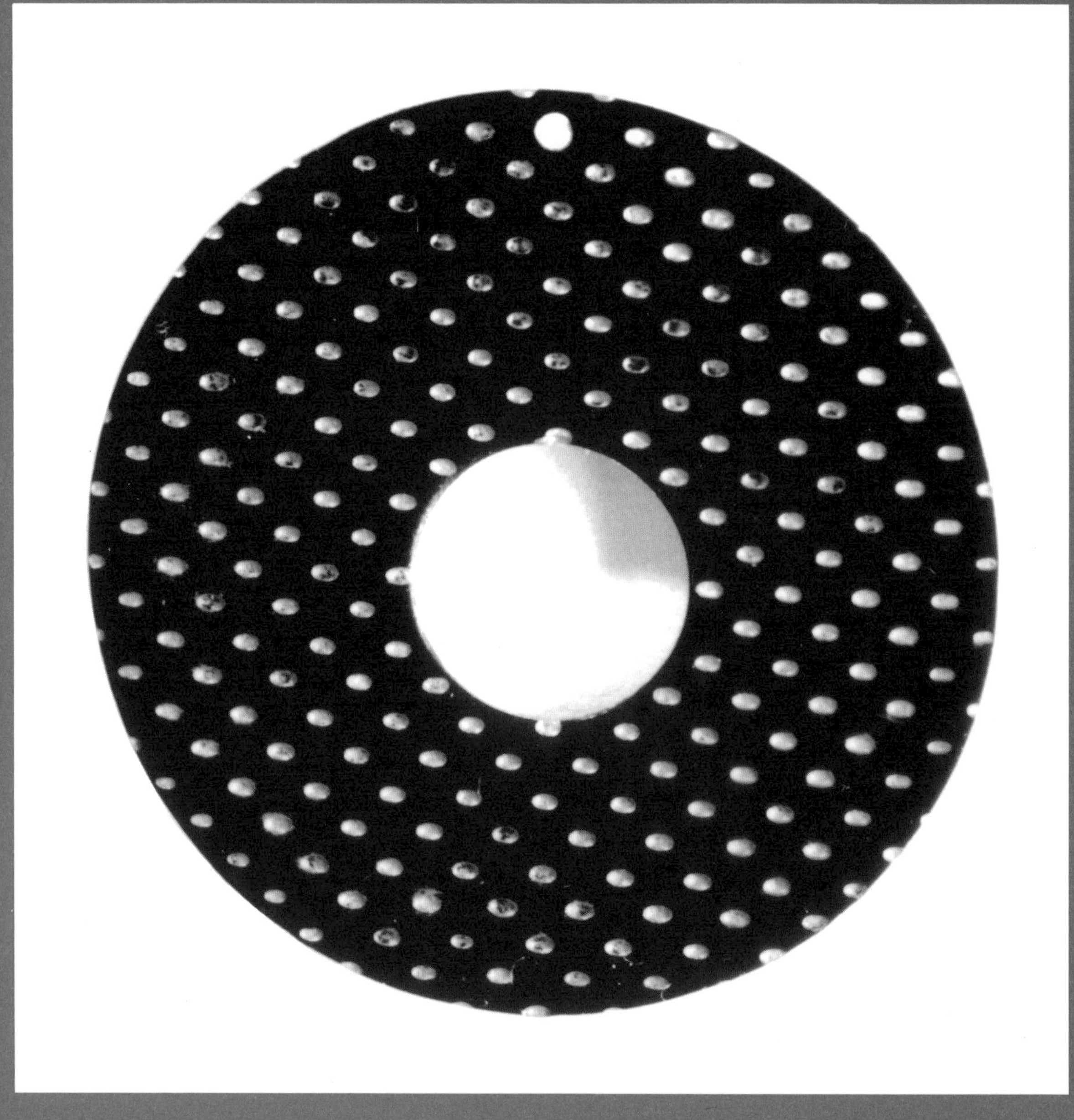

Kris Patzlaff

Depletion Gilding on Tumbaga Alloys

As a means of acquiring a gold-enriched surface, tumbaga and the depletion gilding process offer a viable alternative to traditional mercury gilding, gold leaf and electroplating. In this process heat is used to separate constituents of an alloy into precious and base metals, which are isolated through their oxides. Acids called pickles are then used to dissolve the base metal oxides, leaving a surface layer of the pure precious metal. Depletion gilding of tumbaga alloys is less hazardous than mercury gilding and more durable than gold leaf. It is also less hazardous than plating, and can easily be accomplished in the studio without specialized equipment. Tumbaga offers an affordable solution for the craftsperson who wants the colors of gold or silver without the cost. Tumbaga is less expensive than commercially available alloys and through depletion gilding can provide the same colors at a fraction of the cost. Depletion gilding can be used on 12, 14 and 18K gold and sterling silver to enrich surface colors. In addition depletion gilding can be used to cover firescale.

Historical Background

Tumbaga is a term of Spanish origin for alloys of gold and copper, silver and copper, or gold, silver and copper. Although these alloys vary greatly in composition, the term tumbaga refers to all of them. Pre-Columbian metalsmiths developed tumbaga and processes of treatment that resulted in a gold or silver surface on a finished object. The process by which this surface is accomplished is known as *depletion gilding, depletion silvering* or *mise-en-couleur*. The depletion process uses heat and acid baths or pastes to deplete or remove the base metal from the surface while allowing the precious metal to remain. Through this process pre-Columbian metalsmiths were able to achieve a rich golden surface on an alloy that contained as little as 12% gold by weight. (Lechtman)

At least a millennium before the rise of the Incas, Andean metalsmiths developed tumbaga alloys. Most instrumental in the development of metal technology were the Chavin, Moche, and Chimu metalsmiths. These cultures inhabited the area now known as Peru. Although technical development was purely local, the knowledge of Andean metallurgy was spread from south to north. Arriving fully developed, their techniques reached Ecuador, Columbia, Venezuela, Panama, Costa Rica to the south and as far north as the western area of Oaxaca, Mexico.

Lime Flask, Seated Figure
8¾", cast tumbaga. Quimbaya Style
University Museum,
University of Pennsylvania, Philadelphia

Metalworkers from Columbia and further north preferred to cast tumbaga alloys. Various Indian cultures that cast objects use the cire-perdue or lost wax casting method. Also in use was a mold carved directly in a native black stone. Andean metalsmiths worked principally with sheet metal. Hammered sheet metal was formed into vessels, breastplates, masks, headdresses, jewelry and other objects for ceremonial and personal adornment. Their technology of raising metal was much the same as ours today. Their stakes were carved from hard-woods and often had intricate designs to create a decorative relief on the finished object. The working of sheet metal may have led to the discovery of depletion gilding. (Lechtman) As most of us are aware, because metal hardens when hammered, it must be annealed to regain malleability. When heated, the copper in an alloy reacts with oxygen to form a layer of copper oxide scale. The silver or gold is less readily oxidized. After repeated sequences of the annealing, descaling and hammering that are necessary to form the metal, much of the copper is depleted from the surface, leaving that region rich in silver or gold.

Distilled acids like those used today were not available to Andean metalsmiths. They may have accomplished descaling or pickling by using urine or acetic acid and juices from fruits or leaves or naturally corrosive mineral mixtures. Significant amounts of potassium nitrate, sulfate of potassium, copper, iron and salts were available in the soil. Mixtures of potassium nitrate and salt, or ferric sulfate and salt are as effective as sulfuric acid. (Tushingham) Spanish accounts of metallurgical practices mentioned the use of a "gold medicine" plant, which is believed to have been ban oxalis plant. Objects were rubbed with the plant and then turned to gold. (Emmerich)

Depletion gilding of tumbaga alloys dominated New World technology for nearly two millennia. The commitment to the colors of gold and silver was founded in about 1000 BC with the spread of the Chavin religious cult. (Lechtman) The emphasis on metal technology has strong cultural significance. Unlike Old World metal technology, the Indians' technology did not center on the development of weapons or coinage of precious metals. The highly esteemed silver and gold weren't used as a system of exchange, but rather for the display of social status and religious beliefs. Metals served as a symbolic function, with color being of primary importance.

Both gold and silver figured prominently in mythology. The most innovative aspects of Andean metallurgy arose as a response to achieving the culturally desired colors of gold and silver. (Lechtman) The Inca's believed gold to be the "sweat of the sun" and silver "the tears of the moon." One account of gold's ceremonial importance is the legend of El Dorado from Columbia. Early Spanish accounts read:

> *"They stripped the heir to his skin, and anointed him with a sticky earth on which they placed gold dust so that he was completely covered with the metal. They placed him on a great raft...and at his feet, they placed a great heap of gold and emeralds for him to offer to his god...as the raft left the shore, the music began, with trumpets and flutes... when the raft reached the middle of Lake Guatavita, they raised a banner for silence. The gilded man then made his offerings, throwing out the pile of gold into the middle of the lake...With this ceremony, the new ruler was received, and was recognized as god and king"*
>
> — Juan Rodriguez Freyle (Bray)

Making a Tumbaga Ingot

Making a tumbaga ingot follows the same general principles as the making of any nonferrous metal ingot. The equipment and technique is the same. Refer to standard textbooks for more information if you are not familiar with this process. In making tumbaga alloys it is critical that the atmosphere be as free of oxygen as possible. This is achieved by creating a reducing atmosphere in which the heat source provides an abundance of fuel. This will combine with the available oxygen and prevent it from attaching to the metal. After the ingot is poured and cooled, it is removed from the mold and processed into either sheet or wire.

As mentioned above, the term tumbaga refers to a family of alloys rather than a specific recipe. Experimentation is encouraged, but the following guidelines are given as starting points. Alloys of gold and copper in which gold makes up from 25% to 40% of the mixture (that is 6-9K) will yield successful results. An alloy with as little as 12% gold (3K) cannot be made fully yellow, but can achieve a red gold color through depletion gilding. Alloys of silver and copper containing from 25% to 50% of the precious metal will yield satisfactory results.

The Depletion Gilding Process

Depletion gilding encompasses the processes of heating, pickling and burnishing. These three steps are always performed in sequence (diagram on right), and will be called one cycle.

Heating

Using either a kiln or torch, apply heat to the tumbaga object to oxidize the copper in the alloy. In the first two or three heatings the metal is brought to annealing temperature with subsequent heatings taken only to the point that rainbow colors are seen. Alloys with a high gold or silver content will require fewer heatings than those containing less precious metal.

Pickling

The metal is pickled after each heating to remove the oxides just formed. By depleting the alloy of copper, the surface is left richer in gold or silver. Pickling is done in a strong, hot solution of commercial pickle. Those wishing to more closely duplicate the ancient Andean smiths might want to try any of the following mixtures, but I found little advantage beyond historical interest.

- 50% oxalic acid and 50% salt added to water to make an aqueous paste.
- 50% alum and 50% salt added to water to make an aqueous paste.
- 60% potassium nitrate, 20% alum and 20% salt added to water to make an aqueous paste.

The piece is left in the pickling bath until it's clean of all oxidation and then rinsed with clean water. Plastic tongs are used to remove the metal from pickling baths to protect the surface against scratches.

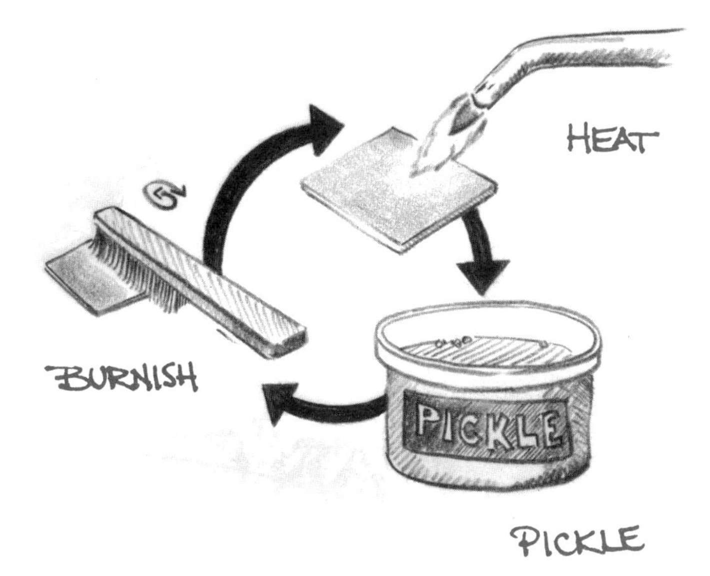

Burnishing

After each pickling and rinsing, the metal surface must be burnished. Without burnishing, the bond between the surface skin and the rest of the alloy is weak, allowing the gold or silver to scratch or chip off. Note that planishing can be substituted for burnishing. Burnishing is usually accomplished with a fine brass wire brush (not the brass brush found in grocery stores), a steel or agate burnisher, or 0000 steel wool. The use of the steel or agate burnisher results in a smooth surface finish and the brass wire brush gives a slightly textured surface. The decision on which to use depends on the nature of the surface and object being worked. A steel burnisher is difficult to use on a piece with small crevices, for instance, but the brass brush will do nicely. The same is true for highly textured surfaces, and of course a combination of the various burnishing tools can be used. The use of steel wool after a steel burnisher or brass brush will soften some of the lines that are produced.

Number of Cycles

The number of cycles required to produce a gold or silver surface varies according to the percentage of gold or silver in the alloy. Those alloys with larger amounts of gold or silver require fewer cycles than alloys with smaller amounts. The silver tumbaga alloys require fewer sequences than gold tumbaga alloys. In a sample of 50% silver and 50% copper, the alloy needs only three heatings to produce a silver surface. A sample of 40% gold and 60% copper needed five cycles to reach a pure gold color. Alloys containing 25% silver or gold require five sequences or more.

The number of cycles is also determined by the color desired. Stopping short of depleting all of the copper from the surface can result in different colors. As mentioned earlier a 12% gold alloy can result in a red gold color. A color resembling 14K red gold can be accomplished with only a few cycles.

Depletion Process for Gold-Silver-Copper Alloys

The process for depletion gilding gold/silver/copper alloys follows basically the same process as the other tumbaga alloys with one additional step. After producing a piece of sheet or wire of the three part alloy, the color of the metal looks silvery with a tinge of yellow. The depletion process is followed as described above to deplete the copper from the surface, resulting in a surface rich in both silver and gold. To produce a gold surface, the silver must be removed. A paste of 80% ferrous sulfate, 20% salt and water is made and applied to the work. The piece is cleaned and checked periodically to determine the surface color. This process can take up to 10 days before a gold surface results.

Thickness of Gold or Silver at the Surface

It's difficult to determine the thickness of the gold or silver layer at the surface without the aid of elaborate laboratory equipment. In the process described above, the gold at the surface is quite thin, suitable for objects that will receive only light handling. Samples that are heated only to low temperatures tend to oxidize slowly, requiring up to 20 cycles, but seem to have a thicker layer of gold at the surface.

Using Tumabaga Alloys and Depletion Gilding Casting

Tumbaga alloys can be used in casting processes. The different constituents are mixed in the crucible when ready to cast. As when making an ingot, use sufficient flux and stir the melt with a carbon rod. After cutting sprues and cleaning the casting, the depletion process is followed as described above.

Fabrication Techniques

Tumbaga responds to most soldering, hammering, shaping and texturing like sterling silver. Of course the melting point of a particular sample will depend on the alloy you've made, so take care when annealing and soldering. Experiments are recommended to determine which grades of solder will be appropriate.

Soldering

Soldering is done before the depletion process. Flux is applied only to the areas that will be soldered, because in this case the oxidation that results from lack of flux is a desired effect. During multiple soldering operations, the piece is heated and pickled several times which of course begins depleting the copper from the surface. It is advantageous to burnish the object between solderings. To protect solder joints, heating for the depletion process is best limited to a lower temperature, heating the metal only until rainbow colors are seen. Note that at these lower temperatures more cycles will be required to achieve a skin of precious metal.

Taking Advantage of the Alloy Components

Exciting possibilities come to mind when you think of the tumbaga as a layered piece of metal. By inventing ways to control the exposure of the alloy itself next to the pure metal surface, a two tone effect becomes relatively simple to achieve. The following example might make this more clear.

Prepare a sheet of a tumbaga alloy, then bring up a pure metal surface by depletion gilding as described above. For this example we'll use an alloy of 25% silver and 75% copper (shibuishi). After 5 or 6 heat-pickle-burnish cycles you have sheet of metal that appears to be silver but is rosy colored in the inside.

This sheet is now roll printed in the usual way. In this example we'll use a template cut from stiff paper (left). The result is a relief that is still all the same silver color.

To create a two-color effect the piece is sanded, using a fine grit paper supported on a rigid backing. This will prevent the abrasive from reaching lower areas created in the roll print (below). By sanding off the enriched skin, the rosy-colored base metal is exposed. This can be left as is or chemically treated to create a patina. In this example a very dilute solution of liver of sulfur will turn the raised areas plum colored while leaving the silver recesses bright.

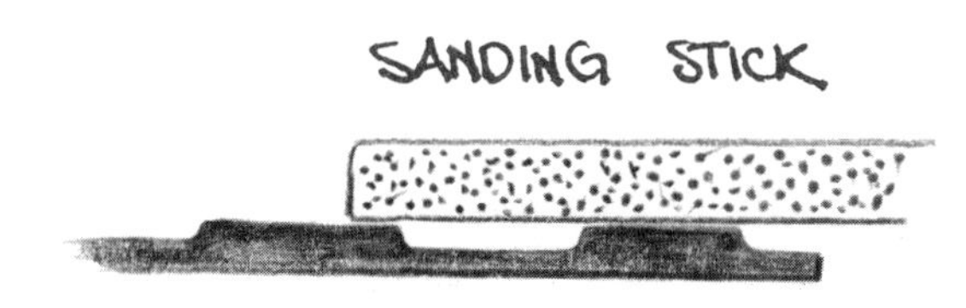

Sandblasting can also be used to achieve a two-tone effect but it is difficult to control the amount of metal being removed. The texture of the sandblasting makes it difficult to see the color of the alloy, but when heat coloring, the sandblasted area will oxidize and the gilded area won't, leaving an interesting contrast of color. Heat coloring results are difficult to control or predict, but then there's something to be said for spontaneity.

Tumbaga alloys are close in composition to the Japanese alloys of kuromi-do and shibuichi, if not actually the same in some instances. It follows then that rokusho patina, a traditional solution used on these alloys will work here. Because rokusho tends to color various alloys differently, care must be taken when sanding to expose the alloy evenly in preparation for patination. Although it might seem to the naked eye that a consistent color had been reached, rokusho treatment will quickly point out discrepancies.

Notes

1. Bray, Warwick; 1979. *Gold of El Dorado.* Harry N. Abrahms, Inc. New York. p. 60.
2. Emmerich, Andre; 1984. *Sweat of the Sun and Tears of the Moon.* Univ. Washington Press. Seattle, Wash. p.187.
3. Lechtman, Heather; 1984. "Pre-Columbian Metallurgy", *Scientific American.* 250:56-63.
4. Tushngham, I.D.; 1976. "Metallurgy"; *Gold for the Gods.* Royal Ontario Museum Exhibition Catalogue. pp. 58-59.

Kris Patzlaff is an independent jewelrymaker and metalsmith who holds an MFA from Southern Illinois University at Carbondale. This material is the result of a research project funded in part by the NEA and SNAG.

Claire Sanford

Patination

The chemical patinas described below are all potentially dangerous if not handled carefully and with the appropriate precautions. Notes throughout the following pages indicate the minimum protection required for each solution and technique. In addition, some general rules apply whenever you are working with patinas:

- chemicals should never be handled directly: wear gloves
- always work in a well ventilated area
- never work where food is prepared or eaten
- know the properties of your chemicals; never use "mystery ingredients"

With few exceptions, the formulas and approaches described can be adjusted to produce a broad range of variations. I am a strong believer in accidental discoveries and encourage experimentation. I have tried to note possible problems and what I know definitely doesn't work. Don't be discouraged; some of the patinas take a sensitive hand and eye and demand some practice to master.

Although most patinas are applied after a piece is complete, some are best done before the work is finished. Because these patinas are heat sensitive, any fabrication done after the coloration must be cold-joined together. It is very important to think through all the steps before you start, particularly if there are elements (i.e. pin stems or earring posts) which you may not want to be patined.

Pat·i·na (pat'e-ne) n. Also pa·tine; pa-ten.
1. A thin layer of corrosion, usually brown or green, that appears on copper or copper alloys, such as bronze, as a result of natural or artificial oxidation. The sheen produced by age and use on any antique surface.

from The American Heritage Dictionary

gloves, well ventilated area

Works on:
Copper, sterling silver, and bronze or brass (very subtle). It does not affect nickel silver.

Colors:
On copper and silver the range is from magenta/blue (hard to maintain) to brownish grey to grey to black. On brass and bronze the only color is a subtle gold. The patina is a very thin sulfide layer, making this appropriate for textured surfaces. Because the solution works very differently on brass and bronze than it does on copper or silver, it can be used to highlight of silver-with-brass or copper-with-brass combinations.

Directions:
For 2 to 4 cups of hot water add a chunk of liver of sulfur the size of a pea. Dissolve completely, test and adjust the strength of the solution. This will vary according to the age of the chemical and the color desired. Make sure the metal you wish to color is perfectly clean and free of any oil. Wearing gloves, hold the piece by the edges or suspend it by a wire and dip into the warm solution until the color is achieved. A dark patina is best built up slowly by alternately dipping and rinsing in warm running water. If the surface is a little uneven try scrubbing very lightly with a Scotch-brite pad, rinsing and dipping again. This works especially well if the metal is textured. For small areas, the solution can be applied with a brush, but be careful when rinsing that streaking doesn't occur.

For coloring brass and bronze, use a very very very dilute solution and dip/rinse as described above. The results will be subtle. If you don't get any color change, keep diluting with warm water; even though it sounds illogical, this often works.

Notes:
If the solution is too strong the color will develop almost instantly and form a precipitate on the surface that rubs off easily and leaves an uneven surface. Spotty coloring can also happen if the metal is not perfectly clean.

I prefer to use liver of sulfur in lump form rather than concentrated liquid but this is only a personal preference. Both have a limited shelf life once opened and should always be stored in a dry dark place. Once the solution is mixed up it has a fairly short life so it's best to dispose of it if not re-used within a day or so. Store in an opaque container in a dark place. If you notice a white powdery crust on the surface of your lump liver on sulfur this means it is getting old and losing its effectiveness. Use more of the chemical when mixing up the solution.

Related chemicals include barium sulfide and ammonium sulfide. These work in a similar way but produce slightly different color tones.

Finishing:
When the desired color is achieved, dry thoroughly and wax. Rub a small amount of wax onto the surface and buff lightly with a soft cloth. If the metal is heavily textured, use matt acrylic spray (see Resources).

COPPER SULFATE (aka "Green Patina")

Gloves, good ventilation.

Works on:
Copper, brass and bronze (not much difference between the three). Use this when a green color is desired but the piece cannot be heated.

Colors:
Pale, variegated, crusty green color. This patina is pretty thick and opaque so textural details are often lost.

Directions:
After much experimentation I have found that the commercially prepared "green patina" solutions are more effective than copper sulfate alone. There seems to be a "secret ingredient" in the commercial solutions and each manufacturer's recipe seems to have its own particular quirks and characteristics. In general, this process works best when the piece to be colored is grease free and has been sanded to provide a tooth on the surface. Dip, brush or spray the solution in thin layers. Let each layer dry completely before repeating. I've found that with a few of the solutions, drying in the sun helps to develop a better color. Once the desired color is achieved, let the piece dry for another day before finishing.

Notes:
One of the big drawbacks of this patina is its tendency to flake off, making it less desirable for jewelry or objects that will be handled. If this starts to happen during the coloring process, brush the flaking area lightly to remove what would fall off anyway and continue the coloration.

Finishing:
After drying the piece completely, wax carefully or spray with a matt acrylic spray (see Resources). Waxing won't work well if the patina is flaky. The acrylic spray can help to glue down loose bits but the surface will remain somewhat delicate.

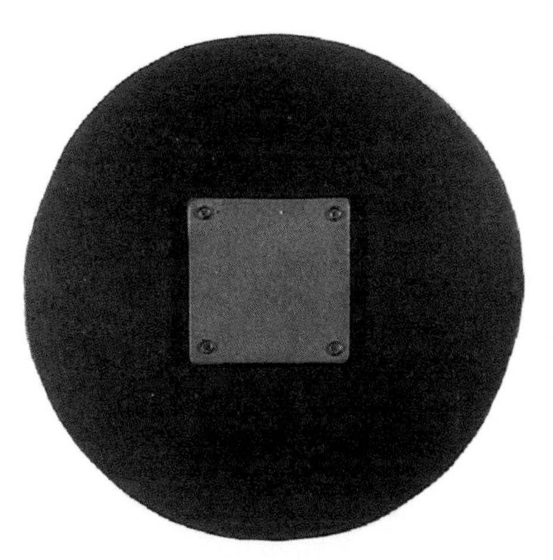

Left, top
Liver of sulfur (outside) and heat patina on copper (center) (made separately and riveted together)

Left, bottom
Dilute liver of sulfur on red brass with sterling (marriage of metal)

Cupric Nitrate

Gloves, glasses, vent hood (or do this outside), respirator

Works on:
copper, brass and bronze
(no difference between these)

Colors:
blue/green with a smooth to somewhat crusty texture.

Health notes:
This can be a very hazardous process if not handled carefully. It is essential that gloves and a certified respirator be worn when doing the patination. Goggles are also imperative because the mucous membranes in the eyes are sensitive to the fumes. The respirator should be fitted with cartridges for acid gasses and must fit your face correctly. The area you work in should also be safeguarded to prevent contamination. Isolate any firebricks you work on and NEVER use them for soldering. At the very least put down a large piece of thin plywood over the work area and ALWAYS clean up the area thoroughly when you are finished.

When using a spray bottle to apply the chemical, the overspray can be a problem. I use a collapsible, reusable cardboard booth that I made to prevent the overspray from getting onto my soldering area. The size will depend on your particular needs. The booth can be made of heavy cardboard or chip board, and hinged at the seams with a couple layers of wide tape. The smaller illustrations show several of the ways the box can be set up with a fan to exhaust gases outdoors. These systems do not need to be elaborate and will repay the few minutes it takes to set them up. I cannot stress enough the need for adequate protection and good ventilation. PLEASE follow these instructions!

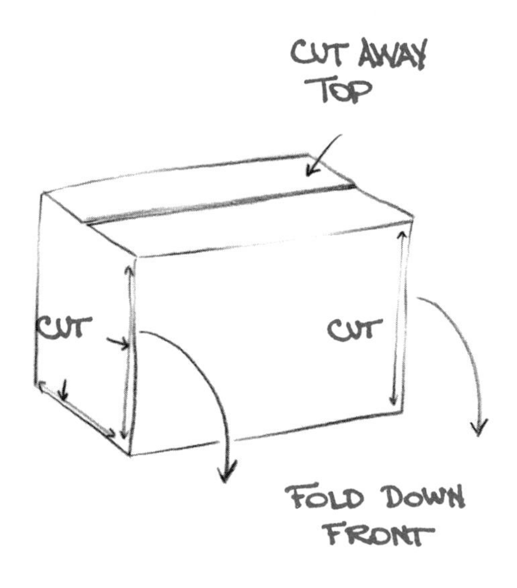

Though simple, an exhaust hood made from a cardboard box can be effective. It folds up for storage between uses.

Here are at least three ways the box can be rigged with a forced movement of air. Use your common sense and test a few options with a smoky bit of rag to find the arrangement that works best for you.

Directions:
Mix 150–200 grams of cupric nitrate crystals in 1 quart of warm water until they dissolve completely. The patination is achieved by a combination of warming the metal with a torch and applying the chemical with a spray bottle or brush. This is repeated until the desired color has been built up. Choice of application method depends mostly on the size of the piece that is to be colored. Small or thin objects do better when the chemical is brushed on whereas larger pieces and thicker gauges color more evenly when the cupric nitrate is sprayed on in a light mist. Very tiny objects or pieces that have a high relief texture are difficult to color evenly because they are hard to heat evenly. Using a large tip and a soft flame (I use a # 5 or 6 tip on a Prestolite acetylene torch), warm the work gently and as evenly as possible. Spray or brush on a light coat of the cupric nitrate solution and dry by heating very gently until the moisture has just evaporated. Repeat several times until a consistent color has been built up. Finish by making sure that there aren't any damp spots left.

Words of caution: don't stop in the middle of the coloring process. It's almost impossible to go back in and add more layers or touch up spots once the piece has cooled down. Be careful not to start applying the solution until the metal has warmed up enough. It takes PRACTICE to get to know when the metal is too hot or too cold to work. If the metal is too cold, the solution will puddle and pool on the surface. Do not let this build up or the patined surface will be very crusty and flaky. Wipe the surface with a paper towel, then dry lightly with the torch. If the metal is too hot, the solution will sputter and turn black immediately. Let the piece cool a little and proceed as described.

The color can by played with to vary it from the familiar bright blue to more green tones. Once a good consistent layer of the chemical has been laid down, warm the piece gently to bring out the greens and some green-brown highlights. Provided you don't burn the surface, you can always spray or brush on more cupric nitrate if you aren't happy with the color. If the patina turns black (which can happen VERY easily) the work must be wet sanded or scrubbed down to the bare metal and the process started again from the beginning.

To color a large object, heat and color in overlapping patches 4–6" square (roughly the size of the pattern of the spray bottle) and, unless you can keep the whole thing warm, try to get the full color in each overlapping patch before moving on to the next.

When the desired color is reached, let the piece cool then rinse it lightly, brushing off any excess chemical or crusty bits with your fingers or a soft toothbrush. Dry the piece as much as possible with a soft towel and let sit overnight until completely dry. Wax or coat with matt acrylic spray (see Resources).

Cupric nitrate on copper
(center bars: cupric nitrate removed by sanding)

Notes:
There are many things that can go wrong with this patination process but don't get discouraged. Things to watch out for include:

- overheating which is easy to identify

- underheating which, especially if the first layers are not "baked on" well enough, can result in the patina popping off as you try to heat the metal

- too much solution being sprayed or brushed on, which will result in crusty black bits forming on the surface

- spottiness occurs when the spray bottle isn't adjusted to the finest mist possible

- it is very important to make sure that all the applied solution is dried with the torch: wet spots will stay on the surface and never really blend in

Variations:
Before waxing or finishing, the color can be manipulated by dipping briefly in a liver of sulfur solution to create deep olive tones. The longer the dipping, the deeper the color. Take care that it does not get too dark! Rinse thoroughly to stop the coloring process, dry and finish as described above.

The mixed cupric nitrate solution will keep indefinitely if stored in a dark, cool place.

Black on Brass

Gloves, good ventilation, respirator

Works on:
Brass only. It does not effect copper, silver, nickel or gold. This patina is especially beautiful when used on brass and silver marriage of metal .

Colors:
Deep, rich black.

Directions:

- ½ cup non-sudsy household ammonia
- ½ cup copper carbonate powder
- 1 cup hot water

Combine the above in a Pyrex dish, or a similar nonmetal vessel that can be placed on a hotplate. A crock pot also works well, but not one you'll ever use again in the kitchen. The copper carbonate will not dissolve in the solution and becomes a sludge on the bottom of the dish. Heat to just below a simmer but DO NOT boil. Maintain this low heat while the work is being colored. The brass must be immaculately clean and free of any copper scale for this to work correctly. Immerse the metal in the bath, either by suspending it into the solution or by laying it in the dish face up. If laying the work in the bottom of the dish, agitate the solution by gently rocking the dish to ensure even coloring. Check after a minute or so to see if the color has developed. Repeat as necessary to get a smooth deep black. Rinse, dry gently with a soft cloth and wax.

Notes:
This process will tell you immediately if there is any grease or oxidation on the surface of the metal. If spots appear that aren't accepting the patina, the surface must be re-sanded. The coloration time depends on the freshness and temperature of the solution. If the color stays a dark olive after 2 or 3 minutes, the solution is either too cold or is becoming exhausted. Try adding a little more ammonia and copper carbonate powder, especially if the solution has been used a lot. This patina does not work well on surfaces that are heavily textured or cannot be completely sanded prior to coloring. You can, however, exploit the color difference between freshly sanded areas and those left with the copper scale.

Black on brass patina, on brass with sterling (marriage of metal)

FUMED AMMONIA

Gloves, good ventilation

Works on:
Best on copper, will also color brass and bronze.

Colors:
Dark olive brown with, if using salt, mottled bright blue and some speckles of the bare metal color (more subtle with brass and bronze).

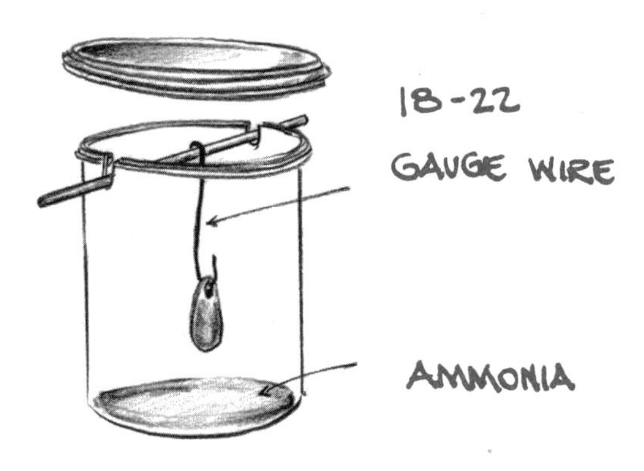

A plastic drum makes a useful fuming chamber. Non-sudsing household ammonia covers the floor of the tub.

Directions:
Suspend a wire across a lidded plastic container which is quite a bit larger than the object to be colored. This can be done by drilling small holes near the top of the container and running a heavy wire across. Hang the object to be colored from the wire so it is not touching the bottom of the container. The metal should be cleaned carefully and given a sanded surface (the rougher the better). Pour a small puddle of non-sudsy household ammonia into the bottom of the container and close the lid tightly. By itself, ammonia will color copper a dark olive color. Bright and dark blue mottled tones can be achieved by wetting the surface of the metal and sprinkling it with salt before fuming. Check after a couple of hours, rinse and re-salt to build up color. Finish with a final fuming of plain ammonia if a darker background is desired. Rinse, let dry completely and wax or spray with matt Krylon.

If the piece that is to be colored cannot be suspended, prop it up on something that prevents it from sitting in the puddle of ammonia (an overturned plastic lid works well) or have the ammonia isolated in a low dish. For large pieces, a jerryrigged plastic tent can also be arranged, using sticks or dowels and a garbage bag. Your particular needs and resources will determine the solution that works best for you. Use a transparent material if possible to allow for viewing of the process.

Notes:
The really bright blue color shows up only after the piece has dried completely. When checking color development it may be necessary to dry the piece before placing it back in the container.

This patina has a wide range of results depending on the length of fuming and the application of salt. The results can be really beautiful, justifying any time taken to experiment and explore.

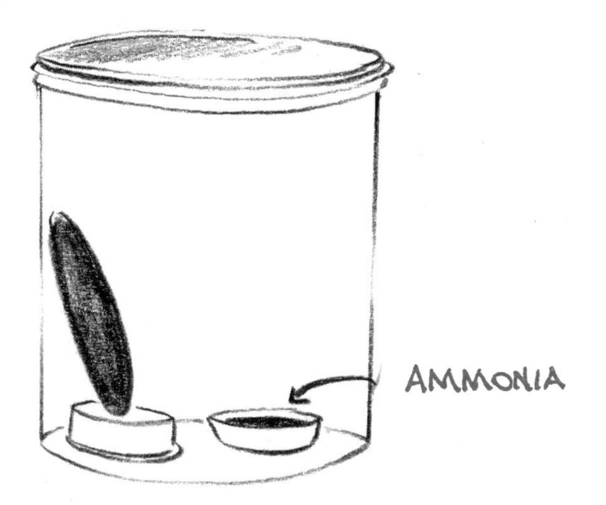

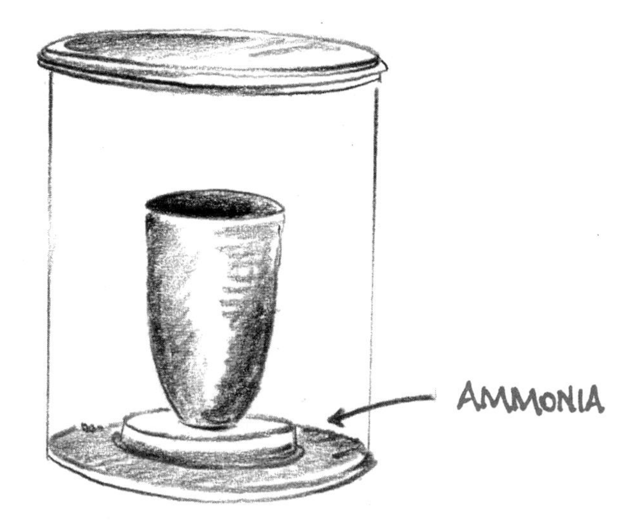

PLASTIC DRUM WITH LID

Alternate fuming arrangements.
A tent is useful for large or irregular pieces.
If possible, use clear plastic
so you can see what's happening
without opening the container.

Sawdust Patina

Gloves, good ventilation

Works on:
Brass and copper. Also very subtle results with bronze and nickel. Each colors quite differently.

Colors:
Mottled gold, green/blue, brown and black on brass. Mottled pink, bright blue, and black on copper. A more subtle mottling on bronze and very subtle markings on nickel. Some etching will occur on the surface of brass and copper.

Note:
This is one of the many different solutions that can be used with sawdust. See *The Colouring, Bronzing and Patination of Metals* by Hughes and Rowe for other recipes. The following recipe has been adapted from a Hughes & Rowe formula.

Directions:

- 16 grams ammonium chloride (optional)
- 16 grams sodium chloride (salt)
- 300 ml non-sudsy household ammonia (double the ammonia if coloring copper)
- 700 ml water

Mix all the ingredients until dissolved completely. This will keep indefinitely if kept in a tightly sealed container in a cool dark place.

The solution can be used with a wide variety of vehicles, not just sawdust. Suggestions include dried leaves, grass clippings, and kitty litter. Just about anything that will hold moisture in suspension and won't dissolve during the patination process is worth trying. The results can be very different and wonderfully unexpected, so spend a little time experimenting. When using sawdust, note that it is the coarseness that gives the patina its pattern. Also, avoid sawdust from plywood or pressboard; the glue they contain can interfere with coloring.

Moisten the sawdust with the solution so that it is damp but not soggy. Place in an airtight plastic container and completely bury the piece to be colored. Leave some scraps of the same metal you are coloring near the top so they can be checked periodically without disturbing the work. As a rule, the finer the sawdust, the less you want to disturb the work during the coloring process.

Sawdust patina on copper with sterling (riveted); shorter exposure, coarser sawdust

In sawdust patina, the work is buried in dampened sawdust or similar materials. Any tightly sealed container can be used.

Coloring time varies a lot depending on the metal and type of vehicle used. For brass allow 12 to 24 hours (the finer the sawdust, the longer the time). Copper colors faster; 4 to 6 hours is usually sufficient but the piece can be left in longer.

Grass clippings or leaves aren't as absorbent as sawdust and require more solution to get the coloring reaction going. A larger quantity should be prepared to allow for the extra dampness in the bottom of the container. Place the work closer to the top for better coloration, checking and turning the work every hour to get a more even patterning.

When coloring is complete, rinse and dry overnight. Finish with wax or matt acrylic spray (see Resources). Once the solution has been mixed with the sawdust it will not keep more than a few days. It's best to mix it fresh each time.

Notes:
This is a very flexible patina, so don't be afraid to experiment with different amounts of ammonia or salt and the length of time in the sawdust. If the sawdust isn't damp enough the results will appear weak. If the sawdust is too damp, the mottling can't take place and the surface of the metal will just etch away. And remember, the finer textured materials need to be left undisturbed whereas the coarser materials, such as leaves, require frequent turning to color evenly. EXPERIMENT!

Sawdust patina on copper with sterling (riveted); longer exposure, finer sawdust

Finishing Products:

Renaissance Wax works very well on all the patinas with the possible exception of some of the commercial "green patinas." These have a tendency to be delicate and can flake with the rubbing required of the wax application. Other waxes (such as Butcher's Wax) can be used, provided that they don't contain any stripping or anti-tarnish agents. Use a very small amount of the wax, rubbing any excess off with a soft cloth. Buff lightly, especially those patinas that have a smooth surface.

Renaissance Wax is available from:

Light Impressions, Inc
Rochester, NY
(800) 828-6216

Krylon Acrylic Spray Matt Finish (no. 1311) is a good alternative to wax. It is available from most art supply stores and from some hardware dealers. Do not substitute Crystal Clear Krylon or Workable Fixative! Spray on light, even coats (2 should do, holding the can 12" away from the work). Do this in a well vented area!

Chemical Supplier:

Bryant Lab
1101 Fifth St.
Berkeley, CA 94710
(510) 526-3141

Bryant Labs have good prices and a very complete list of chemicals. They also have several books on patination. Call for a price list.

Respirators:

I am not referring here to a disposable paper mask. These are made to catch particles and are not effective for the fumes described in this article. Be absolutely certain that the mask you get fits well and has the proper cartridges for the work you are doing. Women should take particular note because most respirators sold in paint supply stores are medium- or large-size which won't cover a thinner or smaller face completely. Smaller masks are made and can be found in safety equipment catalogs. Attention to this can help prevent some serious health problems.

The proper cartridges must protect for ACID GASSES. Don't think that cartridges that protect for paint fumes and solvents will work. Change the cartridges frequently; most are only good for a few "breathing hours." It is very important to keep your respirator tightly wrapped in a plastic bag when not in use!

Recommended Books

The Coulouring, Bronzing and Patination of Metals, Richard Hughes and Michael Rowe, 1982 & 1991. Watson-Guptill Publications, New York

Contemporary Patination, Roland Young, 1989. Sculpt-Nouveau, San Rafael, CA

The Art of Patinas for Bronze, Michael S. Edge, 1990. Michael S. Edge, Springfield, OR

Claire Sanford is an independent jeweler and metalsmith whose work appears in many collections. She lives outside of Boston and is a partner of Top Dog Studios.

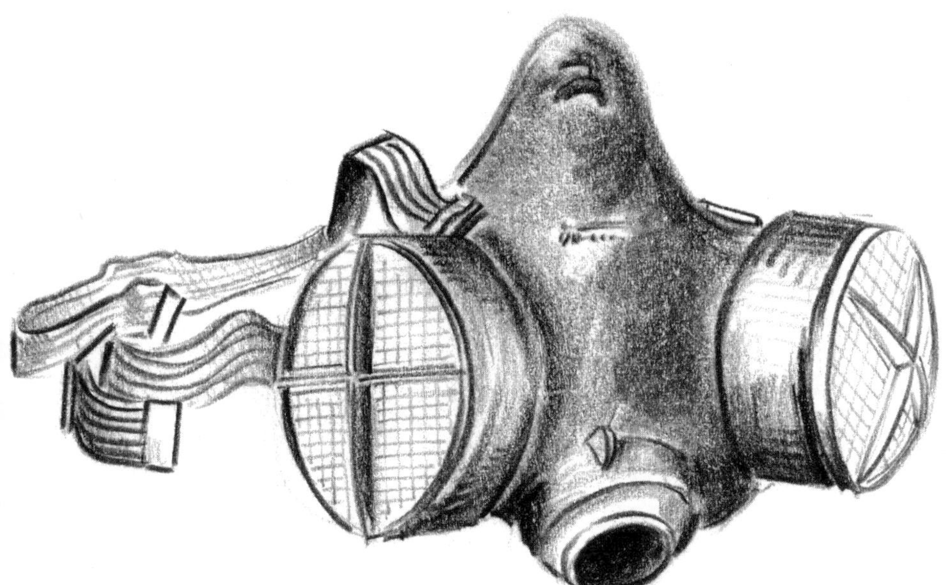

The purchase of a top quality respirator is money well spent. Be sure to use the appropriate filters, and seal them in a plastic bag when not in use to extend their life.

Mary Ann Scherr

The Instant Etch Process

As artists we tend to see ourselves as being in the inspiration business. Many artists believe that talent will get the work done, and that the effort they put into the job is along the lines of facilitating that talent. Edison's famous observation that invention is 10% inspiration and 90% perspiration seems clever, but perhaps not always appropriate to the artist.

For many years I've made work that depends upon graphic surface patterns, and for as many years have accepted the frustrating time, labor and equipment requirements inherent in the photo-silk screen etch process. My current work with computer graphics adds yet another level to my desire for refinement and etch integrities. Starting from this specific frustration, I have been developing a system that brings small scale silk screen resist etching to a new level of efficiency. Besides making my efforts remarkably faster, this process has taught me a lesson about the way ideas grow and the value of collaboration.

A chance exposure to a vendor's demonstration at a trade show brings Edison's comment into focus. While watching the demo, that sudden, inspired, intuitive "click" revealed a long-sought solution. What I saw demonstrated was the Color Printer®, a system for printing designs on cards and stationary. Working with several suppliers and scores of students, I began what would become a 3½ year research and development project that would adapt this machine to reproduce multiple images in acid resists on metal. The Instant Printer® commands only minutes of effort compared to the usual hours or days needed to complete the photoetch process as it has been done with photosensitive materials. I am indebted to the many students at Duke University and the Penland School for their willing help as I experimented to refine the process and locate nontoxic resists and solvents.

The procedure and equipment are now commercially available. I trust that it will be refined, enlarged and improved upon, and that it will be put to uses I have not imagined. This process is a tiny idea in the scheme of things, but through it I have come to see that this is the way things get done, through small ideas that are given form through collaboration and effort.

Abstract

The silk screen Instant Etch® process for reproducing single or multiple images on metal differs from most of the current photo silk screen processes being used. The Instant Etch system requires very little equipment or time and uses non-toxic materials to produce delicate, bold and complex images on the surface of almost any metal.

Working from images produced on a standard copy machine, this process can reproduce original line art, dot-generated photography, patterns, textures and illustrations as surface embellishments on metal. The desired image is converted into a silk screen that deposits resist on the metal. After the resist is dry, the piece is etched with acid or some other corrosive compound. The process usually takes less than one hour from paper art to metal art, depending on the time required for etching.

Using a nontoxic screen cleaner, the silk screen may be reused for duplicating designs. With appropriate care, screens will yield up to 20 prints. My own experience is with the Print Gocco B6 model manufactured by Riso Kaguku Corporation. See endnotes for suppliers. I am aware that other companies make similar products, and it may be possible for the enterprising metalsmith to build his or her own frame and light fixture. The description that follows refers specifically to this equipment, but the information can be adapted for other configurations.

Process

Make a copy of the desired image with a standard paper copier. The art may be drawn with ink or reproduced from a printed image, for instance an illustration in a book. Remember that copyright restrictions may apply and permission should be obtained. If reproducing a photograph, lay a dot screen (eg Letracopy) over the photo before copying it. This will convert the grey tones to a dot pattern.

The copy must be a carbon print of medium black density. Too dense a reproduction (shiny black) will cause the carbon to stick to the screen. The image must fit within the glass window of the press with at least ¼ inch margin around it. The press is available in two sizes: 4 x 6" and 6 x 9". As shown, secure the press to the table top with tape. The process can only become more difficult if the equipment is sliding around.

Prepare to reproduce the image by inserting AA batteries into their housing and flash bulbs into their reflectors. Place the blue filter against the window to protect the surface of the glass from flash damage. A frosted or scarred window may affect the clarity of the image. As is true in graphic reproduction, marks made with this same blue color will not reproduce and can therefore be used for alignment or notations. Appropriate pens and colored pencils can be bought at many art and office supply stores as "non-reproduceable blue."

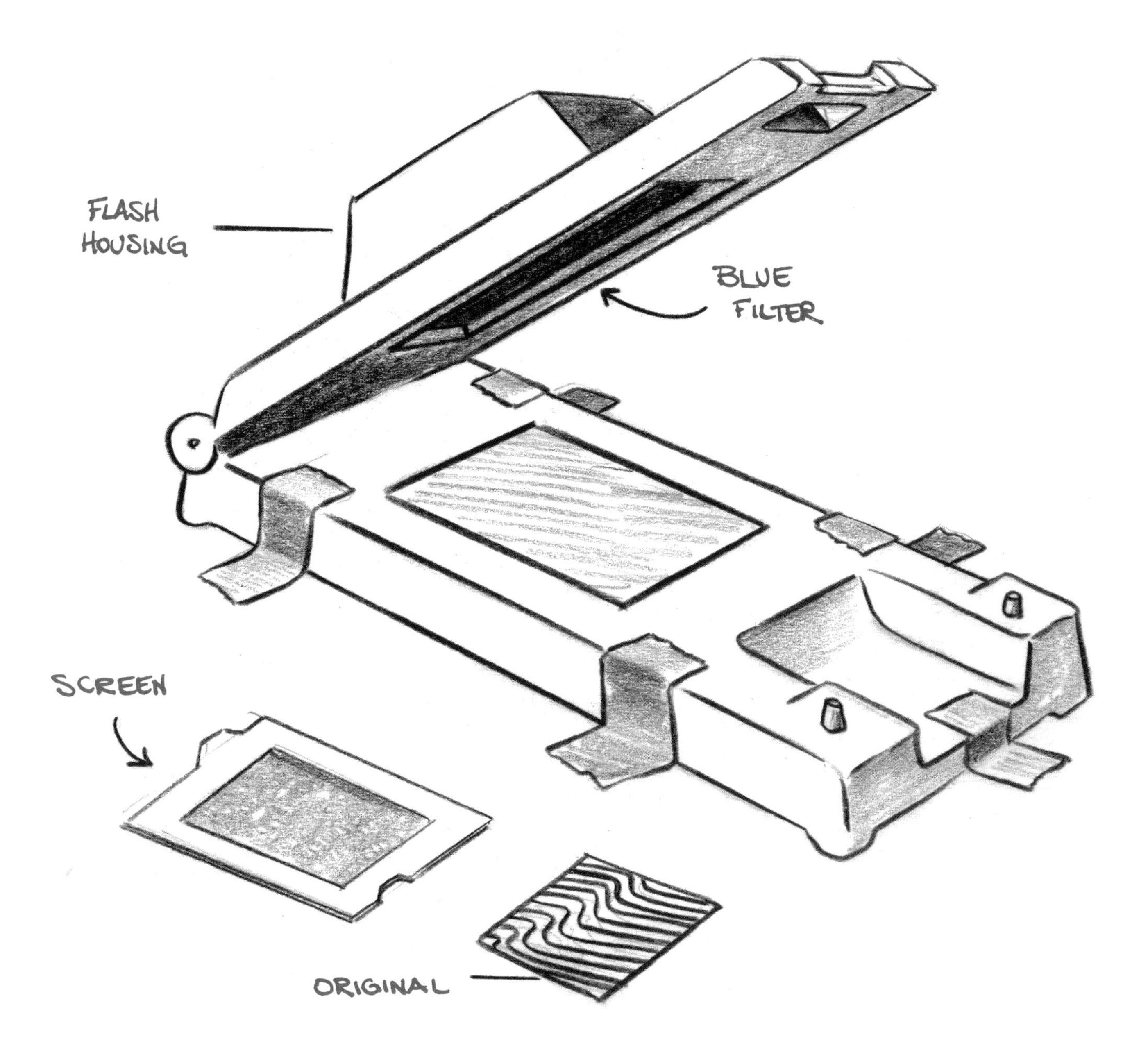

The Riso printer and accessories.

Each image you make will require a Master Silk Screen, which is a piece of fabric stretched into a cardboard frame. Several come with the unit when purchased, and more can be bought as needed. To protect the fine fabric, the screen is covered with a thin sheet of plastic. Remove this cover just before you are ready to print. Bend the screen slightly and place it into the window slots with its arrow facing down and toward the user. Tape the silk screen to the lid to prevent it from moving.

Place the lamp housing into the printer. Position the image in the press and secure it with tape at the top, middle and bottom. Avoid covering the image area with tape and do not restrict the mobility of the pad. When ready to print, the assembly will include the layers shown below.

Press the handle down firmly to trigger the flash bulbs. If you like to be dizzy and temporarily blinded, be sure to look directly at the bulb. Immediately lift the handle with a quick even motion; hesitation will result in an uneven screen image. Carbon deposits left on the screen may be gently removed with a Q-Tip dipped in turpentine. Remove and discard the flashbulbs.

The proper sequence of layers when making a master screen:

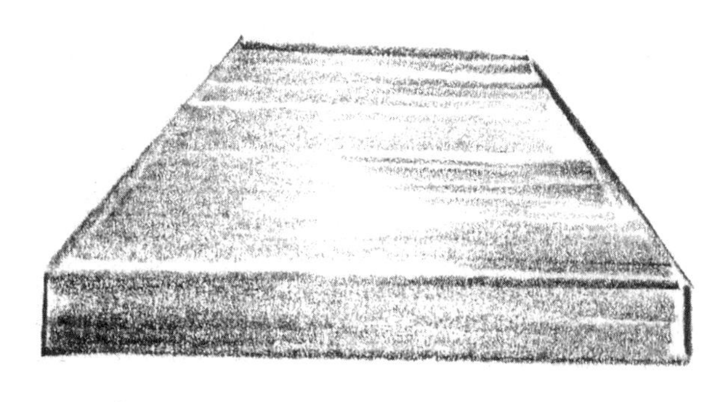

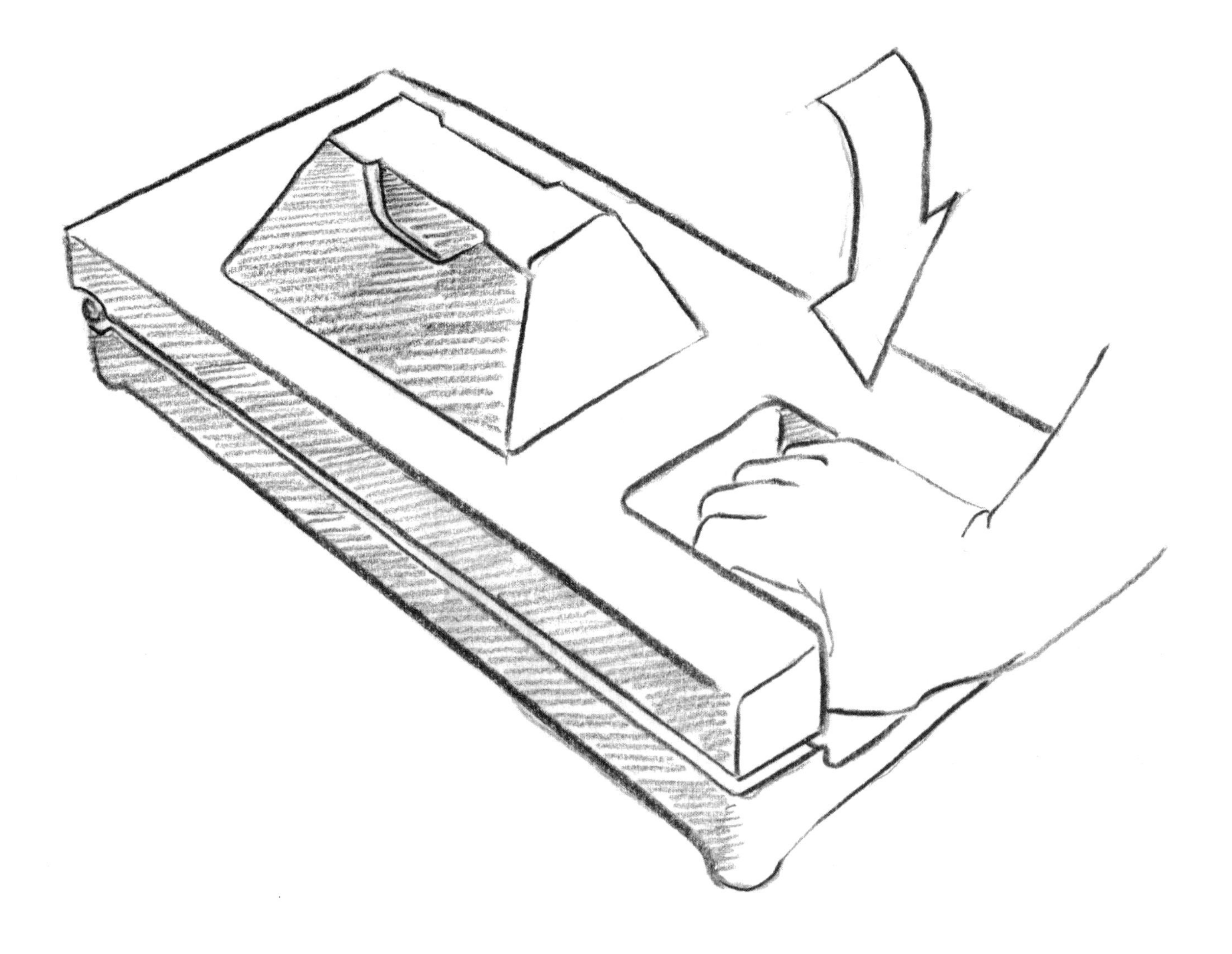

To expose the screen,
press the lid with a firm, brisk motion.

APPLYING THE IMAGE TO METAL

Note:
Though there is very little risk in this process, you might want to wear rubber gloves to protect your hands and keep them clean. Be certain that there is a flow of air through the studio, especially when working is small spaces.

1. Metal surfaces must be scrubbed with non-oily pumice powder. Avoid commercial abrasives, steel wool and scouring compounds. The metal surface must show no water breaks when rinsed. Dry thoroughly with a clean paper towel and do not touch the surface. This cleanliness is critical to the success of the piece.

2. Place the metal on a stable surface and secure it with tape; double sided tape works well for this. Avoid placing the tape where it will get in the way of the image.

3. Register the silk screen image over the metal with its white side up, and make a tape hinge to allow the screen to be raised or lowered onto the metal. To avoid the possibility of the screen shifting, place a small tab at the base of the screen before applying the resist.

4. Spoon enough resist onto the screen to cover the image. Use a squeegee or a firm straight edge (matt board for instance) to pull the puddle of resist across the screen, as shown on page 135. Generally a single firm stroke is sufficient, but the resist can be selectively pushed around as needed to insure complete coverage throughout the image.

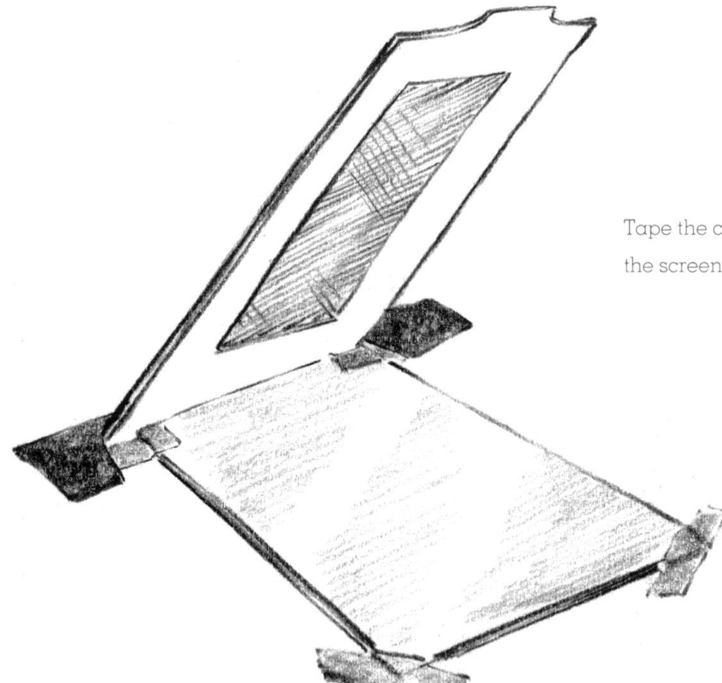

Tape the cleaned metal to the tabletop, then secure the screen into position with a tape hinge.

5. Remove the small lower tape tab and lift the screen gently and evenly. Set the metal aside in a safe place and clean the screen immediately by placing it in a dish of paint thinner. Because of the unpleasant fumes, I recommend a lid and ventilation. Allow the screen to soak for a few minutes to loosen the resist, then clean it with a soft brush. If more than a light scrub seems necessary, let the screen soak a little longer. If further printings of this image are desired, blot away excess thinner and let the screens dry before reprinting.

6. When the asphaltum is dry to the touch, correct printing flaws with the "PILOT" Pen (gold only) or with standard asphaltum. Unwanted sections can be scraped off with a metal scribe.

The metal image is now ready to be etched! In a perfect world, the process has taken about 15 minutes.

Spread a spoonful of resist onto the screen and drag it across the window with a squeegee, trying for complete coverage in a single pass.

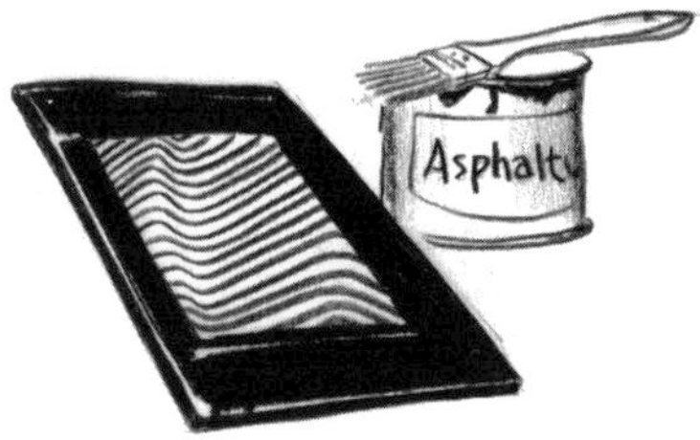

Use asphaltum to mask out other areas of the metal that are not be be etched. Remember to protect the back.

Etching Safety

By its definition, etching involves the use of strong oxidizing agents – acids. Either make up your mind to take this danger seriously or don't get started. Artists have been etching for hundreds of years, so it should be obvious that it's possible to do this without injury. But remember that no one sets out to hurt themselves. Accidents are usually the result of poor planning, insufficient space, or hurried work. By being aware of these dangers and then planning ahead to avoid them, you will make your etching experience safer and more rewarding.

Etching must always be carried out in a well ventilated area. The ideal arrangement has an enclosed, vented fume table. The window is pulled down to seal off the environment and prevent fumes from entering the studio. If your vent system is less sophisticated, you'll be wise to supplement it with a respirator, fitted, of course, with the appropriate filters. Avoid a large overhead fume hood, because it will probably pull the fumes up into your face, rather than protect you from them.

Always wear rubber gloves and a rubber apron when working around acids. Protect your eyes with splashproof goggles, and keep baking soda close at hand to neutralize spills. Acids and any mordant mixtures must be kept in vessels made for scientific use. Keeping acids in peanut butter jars is asking for trouble. Lids must be plastic or glass or they will corrode away. Always mark each container well, stating the name and ingredients of a mixture and the date of its preparation. Never store acids on a high shelf, where someone might accidently pull them down. And always take special care when children or pets are on hand.

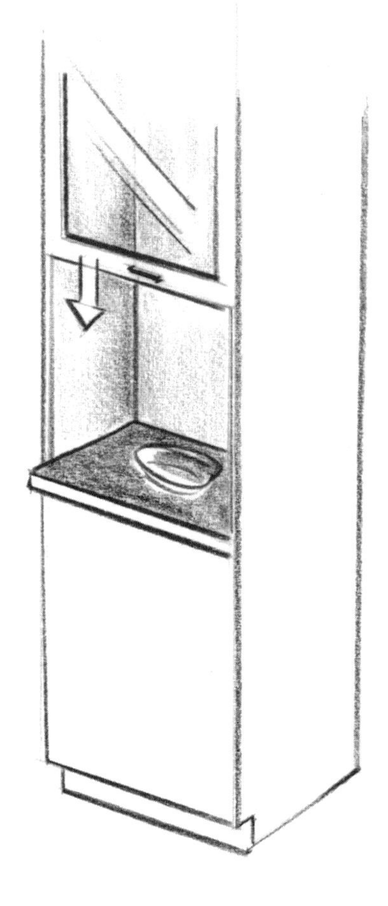

A laboratory ventilation booth allows for a sealed environment when the sliding window is pulled down.

Mordants and Resists

The word mordant refers to any solution used to etch metal. While some acids will attack almost anything, each metal has one or two mordants that are particularly appropriate for it. Resists are the acid-proof coatings that protect some areas of a metal object, and this is of course what makes decoration possible. Again, some resists will stand up to almost anything, but generally each mordant has a few preferred resists. The desired image or the method of creating that image will often dictate the resist as well. Sometimes a photo resist like the one illustrated in the above technique is needed. Other resists are painterly, while still others are especially good at adhering to an irregular surface. Experimentation is needed to determine which mordants and resists you will want to have at your disposal.

Nitric Acid *(NHO_3; also called aqua fortis)*

This is a transparent, fuming liquid that ranges from colorless to yellow. It is extremely corrosive and can become explosive in certain solutions. Since its discovery in 1787, nitric acid has been used as a resist for copper, but the rough bite it usually creates is a drawback when delicate lines are needed. Printmakers often prefer a mixture of hydrochloric acid, potassium chlorate and water, called Dutch Mordant.

For jewelers and metalsmiths, nitric acid is most important as a mordant for sterling silver, where it is mixed with three times its fluid measure of water. That means that one cup of acid would be added to 3 cups of water. It's important to create the mixture in this sequence, always adding the acid to the water. Because the acid is the heavier liquid, it will drop to the bottom of the container and begin mingling with the water. If the process is reversed, the water might sit on the layer of acid, limiting the interface between the two liquids. If the mix activity is confined to a small place, the tremendous heat that is created can cause the acid to splatter.

Nitric acid is also used with three parts hydrochloric acid to create an extremely potent mixture called aqua regia, "Royal Water." This is used to etch gold alloys over 10 karat.

Resists for Nitric Acid

Because of the strength of this mordant, only strong resists will stand up to it. The most common is a tar residue called *asphaltum*. This can be purchased in a refined form through art supply stores. In a coarser form the same stuff is used to spread on driveways as a sealer. Asphaltum is a thick paint that is applied with a brush onto clean metal. It requires from 15 minutes to a couple hours to dry, so plan ahead. A variation is made by mixing equal parts of beeswax and asphaltum. The result, called *hard ground*, is poured onto a slab of stone or steel and gathered up in gloved hands as it cools. It is molded into a ball (like making a snowball) and rubbed on slightly heated metal to create a coating of resist.

Ferric Chloride

($FeCl_3$; also called Iron Perchloride)

This mordant has gained popularity in recent years because of its slow even etch on copper and brass. It is sold as a yellow gravel or, premixed with water, to become liquid of a deep amber color. Ferric chloride is considerably less violent an oxidizer than nitric acid, though of course precautions should still be taken. Of particular value if the fact that ferric chloride is not prone to fuming, so only minimum ventilation is required.

In the oxidation reaction of copper to nitric acid, a hydrogen molecule is released and escapes as a gas. This gives the reaction an effervescence, or a bubbling action that serves to clear away bits of copper that are corroded. The ferric chloride reaction lacks this activity and in fact creates a sludge that can coat the metal and prevent further etching. For this reason, plates are either set upside down or placed in a machine that will spray the plate with acid and allow it to run off.

Resists for Ferric Chloride

One of the reasons for the popularity of this mordant is the wide range of resist materials it can accommodate. Of course the all purpose asphaltum will work, but so will most paints, permanent markers (black is preferred), nail polish, wax pencils, and transfer letters. To stop out (that is, to protect from etching), large areas can be covered with electricians' tape. This is not only quick to apply, but has the advantage of coming off easily as well.

After Etching

The length of time a metal will be exposed to a mordant depends on many factors, including the strength of the acid, its temperature, the size ratio between surface and fluid, and the desired depth of the bite. This is usually checked by "feeling" the etched grooves with a needle. When the bite is complete, the metal is rinsed well in running water, then set to soak in a neutralizing bath. In the case of nitric acid, this consists of a concentrated solution of baking soda in water. For ferric chloride, make up a strong mixture of ammonia and water and allow the metal to sit in it for several minutes. Remember, though, that ammonia blackens brass, so don't allow that metal to sit too long. Of course you'll be excited about your work, but remember to methodically clean up all the tools and containers used during etching and to safely store the chemicals.

Suppliers

Note:
The equipment is sometimes available through art supply stores; check your local sources.

Riso, Inc.
3000 Rosewood Drive, Suite 210
Danvers, MA 01923
800 866-2179

Rio Grande Albuquerque
6901 Washington Aveue NW
Albuquerque, NM 69109
800 545-6566
(called the Etch Press®
or the Multicolor Printer®)

Asphaltum Resist
"Silk Screen" asphaltum
"Standard" asphaltum
C.R. Hill Company
2734 Eleven Mile Road
Berkley, MI 48072
800 521-1221

Mary Ann Scherr is a metalsmith and teacher whose influence has been felt in the field for several decades. She has taught at Kent State University and Parsons, and now lives in Raleigh, NC.

18K

Heikki Seppä

Reticulation

Reticulation is a process performed in the studio by a jeweler to create a decorative veined or wrinkled effect on sheet metal. The process is sufficiently predictable to ensure professional results but is at the same time subject to enough mystery to make it as fascinating the hundredth time as it was the first. The process requires only conventional equipment and very little time. The results are apparent immediately and can find application in jewelry, hollowware and many other objects. This technique uses standard jewelers' pickle and a torch. The only safety factors that must be considered are the commonsense precautions that accompany these tools: guard against splashing the acid solution and use the torch only in an area that has been adequately fireproofed.

Historical Notes

The technique we call reticulation is an internationally accepted process and is called "shamarodok" in Russian and "liekkirypytys, or krymppäys" in Finnish. Evidence indicates that the technique was discovered by accident. Craftsmen like myself worked with it, and continue to expand on its potential.

My own involvement with reticulation can be traced to the studio of the well known goldsmith Carl Fabergé. In 1870, Carl Fabergé, whose Huguenot parents had escaped from France, took over his family's small metalsmithing studio in St. Petersburg (at one time also called Leningrad) at the eastern end of the Gulf of Finland. His skill and elaborate designs earned a reputation for the firm and under his guidance the House of Fabergé grew to become one of the leading design firms of Europe. After nearly three generations of gold and silversmithing, the Fabergé studio was closed in 1917, an unfortunate victim of the Bolshevik Revolution. At that time the house of Fabergé had over 700 workers, many of them Finns. When, in that same year, Finland won her independence from Russia, many Finnish metalsmiths returned to their native country from St. Petersburg. During my training and subsequent work in the precious metal industry, I was fortunate enough to work with some master craftsmen who had worked for Carl Fabergé.

The "working silver," i.e. the standard silver alloy used in manufactured silver objects in Russia, was 820 parts per thousand fine, or 82% silver and 18% copper. This is still the case in Finland, but sterling is becoming more accepted as the standard. The reduced silver content of this alloy lowers its melting point to 1520°F (825°C), or about 100 degrees Fahrenheit lower than sterling. Because the melting points of the parent metal and solder are so close, soldering the 820/1000 alloy calls for more finesse than working with sterling.

Mostly these workmen knew what they were doing, but on occasion a little too much heat, especially during the last soldering, could "reticulate," or wrinkle the surface of the sheet. As is often the case however, these mistakes became the origin for a new process. The reticulation texture was too handsome to ignore, so it was put to use.

Many products were developed for the market, including cigarette cases, eye-glass cases, makeup compacts, cigar cases, hip flasks, and similar small containers (etui). The best reticulation was done on rather thin sheet (0.5 mm), which could still be formed in a mild contour without tearing. Reticulation, as a surface treatment, was established.

When I came to study at Cranbrook Academy of Art in Michigan in 1960, I thought the whole world knew of this seemingly standard texture-producing technique. But I discovered otherwise when a friend from Finland sent me a sheet of the 820/1000 alloy and none of my fellow students had seen it before. With this one silver sheet I demonstrated the technique, not even knowing what to call it. Dick Thomas, the instructor, walked by and asked what I was doing. When I replied that I didn't know what to call it in English he said, "Looks like some kind of reticular forming to me." O.K., Mr. Thomas, thank you. I'll call it reticulation from now on and since May of 1961 I have.

At that time I had neither seen nor heard of reticulation on gold. I considered gold plating the silver alloy to achieve the gold color, but instead decided to risk a gold sheet to the process. There have been a number of scientific examinations of the reticulation process, but the explanations don't satisfy me. I decided that the only way to get to the bottom of the issue was through personal experiments.

My experiments in '63–'64 were slow, because I had no money to purchase a rolling mill to make new sheets from the ingots of gold that I made myself. Nevertheless, I managed to discover a number of approaches to 14K yellow gold reticulation.

My decision to experiment with 18K yellow gold reticulation grew from a mistake I made in writing up a contract. A customer and I had been talking of the wonders of reticulated gold surfaces and I was commissioned to produce a bracelet with reticulated texture. When I wrote up a commission contract I mistakenly entered the material as 18K rather than 14K gold. For years I had thought that this was an impossibility but instead of trying to correct my mistake in the contract, I decided to try reticulating the higher karat gold. I suffered a couple total disasters, but my new power rolling mill renewed the spoiled sheets time and time again. Finally there was a success.

Since then reticulation techniques have been applied to numerous alloys. In 1973 Paulette Myers went so far as to reticulate nickel silver. Her thesis on the subject of reticulation discovered the new process of nickel-silver perforation, which renders the sheet, not in a dendritic rugosity, but very acutely perforated and partially expanded state, like a raised bread. This was published in a book called *Metalsmith Papers*.

Over the last 30 years I have given numerous workshop demonstrations on the subject in various schools all over the North American continent. The process has been described in several books and can probably be considered something of a standard technique in the field. I'm glad to have been a middleman in bringing it from its origins in Russia to North America.

THEORY

Reticulation depends on the creation of a sandwich of layers that melt at slightly different temperatures. This is achieved by systematically developing a surface layer of a nearly pure metal (silver or gold) in a process called depletion gilding or "bringing up the silver." When silver or gold alloys are heated, a firescale layer that contains oxides of copper and silver ($CuO + CuO_2 + AgO$) is formed on the surface of a sheet. Treatment in an acid bath ("pickling") dissolves most of the copper compounds and leaves a skin of pure metal at the surface. With each heating and pickling operation this exterior layer is made thicker and driven deeper, creating a high temperature skin on the relatively low melting alloy.

Note that pickle dissolves copper oxides, but not silver oxides, and not copper as it resides in an alloy. This explains why a sheet of sterling left in the pickle for a long time is not reduced to pure silver. By heating the alloy, the copper is driven to combine with oxygen, forming copper oxides that can be dissolved in pickle. It is for this reason that the best reticulation is achieved with an alloy that has a higher copper content than sterling.

The rich texture of reticulation is not possible on pure metals but can be achieved on many alloys. It seems to work best on sheet about 0.5 mm thick (24 ga. B&S). Even after reticulation the sheet retains much of the ductility and malleability of the original. My preferred alloy for reticulation is 820/1000 Ag (82% fine silver, 18% copper). Sterling silver (925/1000 Ag) may be reticulated but yields only a weak rugosity or wrinkling. The stock can be purchased from some refiners or ingots of this alloy can be made in the studio and rolled into sheet. When making an alloy, premeasure the component metals and start with either one. Feed in the balance, add flux and stir, but guard against overheating. When the alloy is mixed and fluid it is poured into a pre-heated ingot mold. If starting with sterling rather than fine silver, add about 10% copper by weight.

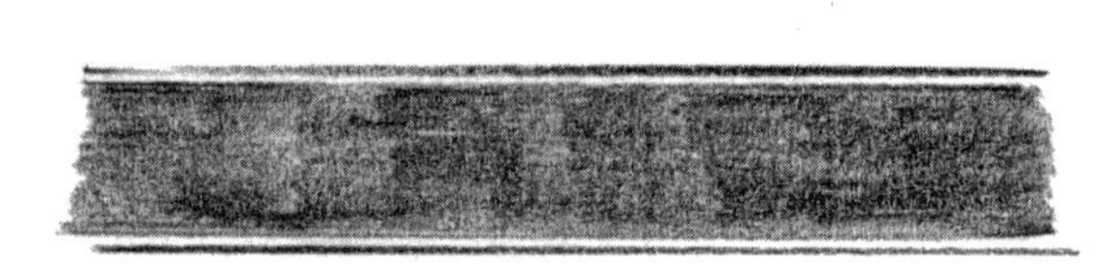

FINE SILVER
PARENT ALLOY
FINE SILVER

By depletion gilding, a fine silver oxide skin is created over a copper-silver alloy. The proportions are for illustration; in fact the skin is much thinner than this.

PROCESS

Anneal the sheet, preferably with a natural gas/air torch, to create a layer of silver and copper oxides. After each heating the metal is cleaned in a freshly mixed Sparex solution or in a 5–8% sulphuric acid pickle. Because copper oxide is soluble in the pickle it leaches away, leaving a fine silver (or fine gold) oxide skin. Scratchbrush both sides of the sheet the first 2 times, using a mild alkaline solution such as a small amount of baking soda mixed with water. Repeat the heat-pickle-rinse-scratchbrush cycle 3 to 5 times. Avoid gouging the surface with tongs, handling the sheet by the edges if possible. Do not exceed 1200°F (650°C).

The purpose of scratchbrushing the sheet is to strengthen the bond between the fine silver skin and the parent alloy. This is ideally done with a slow (800 RPM) brass or nickle silver brush lubricated with very mild alkaline solution. Traditionally scratchbrushing was done with stale beer. This is, of course, bacteria-laden and smelly, but it's a good acid neutralizer. If there are green, purple, or yellow flames bouncing off the silver surface during the next annealing, traces of the pickle were unneutralized. Rinse better next time!

It's important to consider the heating surface before attempting reticulation. The preparation of the heating block (the bed) is important because latent chemicals and moisture may harm the silver sheet. Ideally the bed will be flat, porous and thin. New fiber frax or other non-asbestos boards of about 5 mm thick are ideal. There must be no flux on the board. Heat the bed with a torch or in a kiln. This will not only help in uniform warming of the sheet, but will render the bed porous.

The best reticular growth comes from the use of a rather cool flame. The best flame for silver is a strongly oxidizing mixture. This flame uses more air than normal. It will have a pale blue color and make a hissing noise.

The reticule begins to grow at the heel of the flame, at its underside. Patience pays off. Wait until you have a small area starting to respond to the flame, then maintain that pace by slowly moving the torch forward over the sheet. The clearest indications of proper heating will come from close observation of the surface, which ought to remain dry. If it "sweats" that is the sign of too much heat. Pull the torch back, not sideways, and add more air (or decrease the gas). If the growth-run is interrupted, either by a jerky motion or lack of constant temperature, the wrinkling will stop and cannot be continued without a noticeable blemish. Even a small deviation (⅛ inch) in the distance between the torch tip and the sheet will alter the results. Perhaps this element of risk contributes to the excitement of the process.

The texture will appear at the "heel" of the flame.

Begin reticulation with a medium size flame with airflow so strong that the flame is almost blown out. Work on a strip roughly ½ inch wide to produce controllable results. Regular size soldering torch tips render a path of reticulation from ½"–1" wide. With large annealing torches 2" or 3" paths may be reticulated. The individual reticules may be formed in orderly patterns or the surface may be covered with random wrinkling.

Reticulation is done on an arbitrary size sheet, generally the larger the better. Sometimes during the process the parent metal (the inner layer) in its momentary molten state will ball and leave some areas where the two oxide layers meet. Upon cooling this will usually produce a hole in the sheet, which can be repaired or incorporated into a design. The basic idea is to reticulate a sheet and select choice areas for use, discarding the rest. It is never a good idea to pre-cut sheets "en silhouette" and hope to reticulate them.

Close windows and doors to reduce unwanted drafts. Modern studios often have forced air heating and ventilation systems that create drafts and must be eliminated. The blast of heat leaving the torch appears to be very strong, but any side draft can upset it and ruin the effort. There is only one chance per sheet. Re-reticulation is practically futile.

Hold the torch vertically for your first attempt at reticulation as shown below. Once you get the hang of it, try tilting the torch. Later learn to tilt the angle if desired. Many of my mentors slanted their torches while reticulating. Some of their results were very interesting, exhibiting good dentritic order and unbelievably orchestrated patterns. I have not learned to do that yet.

You'll have better control if you direct the flame vertically onto the sheet. Note that the flame does not touch the sheet.

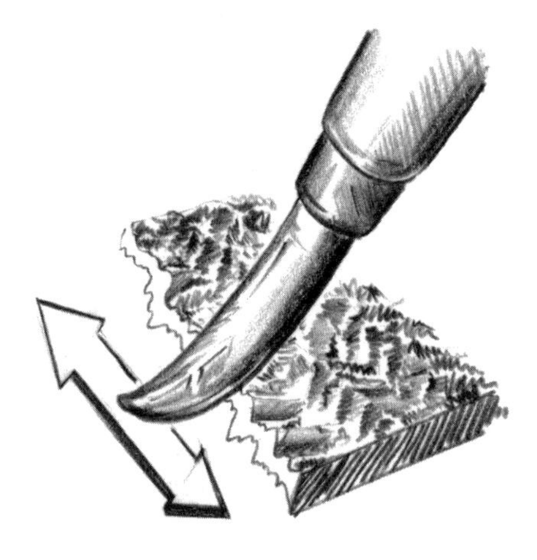

Before soldering an edge, it's a good idea to burnish it like this. Reticulated metal is porous and therefore has a tendency to "soak up" solder.

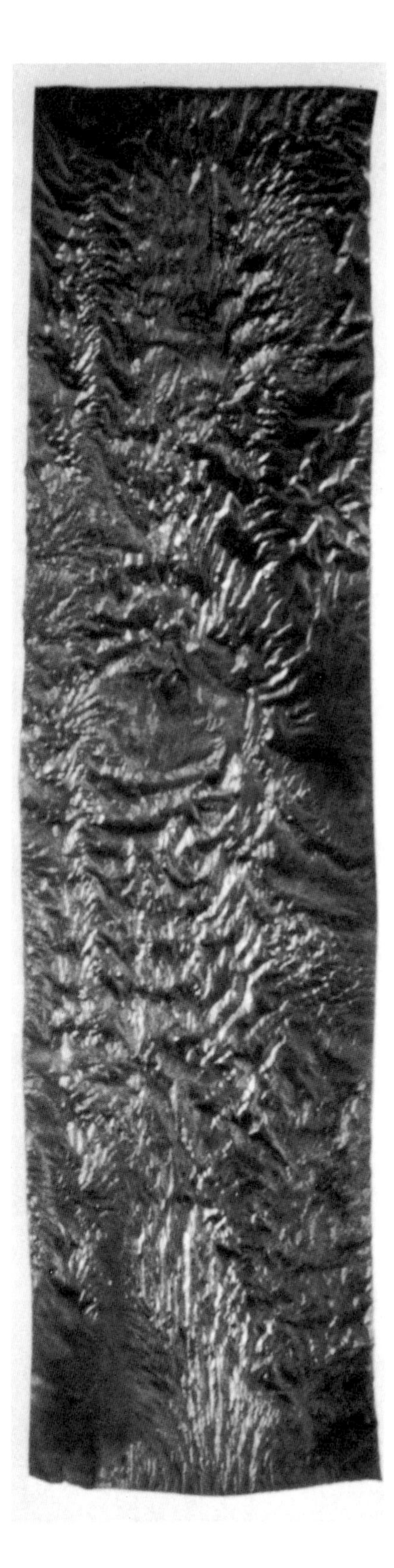

Reticulation can be controlled to make bands or applied in a random pattern

With practice you'll be able to steer the pattern and determine the width of each band, primarily through torch control.

There are several other variations of reticulation control. It is possible, for instance, to leave some areas untextured, reticulating a wonderfully sumptuous path in the middle of a wide smooth strip for a bracelet, or reversing the same, only reticulating edges. By altering the direction of the torch and its height above the metal, distinct rows of texture can be created.

Gouges produced with a scribe prior to the heating process can give some interesting linear growth patterns. Or you might try reticulating just one spot, letting the growth radiate from it. Sometimes centerpunched scutates (dimples) react in an unexpected way. Pre-cut holes and pre-cut patterns slipped under the silver sheet (usually iron sheets or nails) add to the variety of the results.

Gold reticulation in particular requires a natural gas and oxygen torch for the most successful results. 18K need not be annealed and will produce a curious reticular growth but not a dendritic, orderly one. 14K yellow will give a number of results. It needs to be annealed and pickled as described above. As mentioned previously, the gas/oxygen torch must be overladen with oxygen. The recommended thickness for 18K yellow is 0.4 mm (26 ga) and slightly thinner for 14K yellow. Do not judge the results until you have pickled and scratchbrushed the sheet. Then you may be gazing into the possibilities for hours.

Reticulated sheets can be fabricated just like any other jewelry alloy but a few precautions are in order. First, remember that the melting point of Ag 820 is lower than sterling. Avoid the use of hard solder if possible and stay alert when soldering no matter which grade you use. Because the structure of reticulated metal is porous, solder joints are potentially less secure. When joining 2 pieces, say in a T-joint, it is best to file and burnish across the edge to expose some of the parent material for the solder to make a sound bond.

MATERIALS

Reticulation silver (Ag 820/1000).
Stock size is 3" x 8" x .5mm (24 ga).

Available from:

Hauser & Miller Co.
10950 Lin-Valle Dr.
St. Louis, MO 63123
(800) 462-7447

Hoover & Strong, Inc.
10700 Trade Road
Richmond, VA 23236
(800) 759-9997

For pickling silver:
Sparex® (ordinary solution) or 5–8% sulfuric acid in water. Remember to add the acid to the water, not the reverse.

For pickling gold:
Nitric acid from 10%–30% technical grade. Remember to always add acid to water when mixing acids.

Scratchbrushing:
3" diameter 4–6 row fine round wheel, slow speed (800 RPM) Nickel silver or brass or hand operated brushes, but no steel wool. Use mild alkaline solution – baking soda with a few drops of detergent in one gallon water

Heating is best done on a thin piece of a non-asbestos soldering board or a ¼ inch slice hacksawed from a soft fire brick. Do not use charcoal blocks, transite boards, or hard bricks of any kind.

Natural gas is preferred for reticulation, but of course not all communities have natural gas. In that case you should know that propane comes closest, but is somewhat hotter, which will change the reticulation results. Mapp gas is hotter still so special care must be taken. Acetylene/atmosphere torches (eg. Prestolite) are too hot for controlled work on silver, but work okay on brasses and bronzes. Modern Engineering Co. (MECO) St. Louis, MO makes a very good midget torch for natural gas and oxygen use. This versatile torch (tip #4) is excellent for gold reticulation. This torch is available from many catalog supply houses.

Heikki Seppä, a professor for 27 years at Washington University in St. Louis and now professor emeritus there, has pioneered a program of form development through shell structures. He is a noted workshop leader and the author of *Form Emphasis for Metalsmiths*, Kent State University Press, Kent, Ohio, USA.

Glossary

Anneal The process of making metal malleable through the application of heat. In the case of steel, annealing is achieved by a slow cooling after heating the metal to a bright red color.

Anticlastic A form in which the primary axes move in opposite directions. An example of the shape that this creates is a saddle. The opposite is a form in which both primary axes move in the same direction, as in a conventional bowl.

Axial curve This describes an imaginary line moving parallel to a primary axis, along a stake. It is at right angles to the generator axis.

Bi-shell A fabricated object with an interior space that is constructed from two pieces. A clamshell is an example.

Carburization The result of a union of carbon with molecules of a metal at the surface.

Cold connections A large family of joinery techniques that do not require heat. Subcategories of this include chemical connections such as glue, and mechanical connections in which various elements are physically shaped to secure parts together.

Cross peen hammer A hammer with a wedge-shaped face that runs at a right angle to the handle. It is used whenever the flow of metal must be controlled, as in forging.

Crucible A container made of ceramic or graphite used to melt metal for pouring.

Dendrite A systematic crystalline growth pattern.

Die forming A broadly defined process in which a malleable material is pressed into a preconceived form through impact with a rigid device called a die.

Emboss To create a contour in a sheet by pressing a rigid material into it. This usually refers to a relatively small distortion in an otherwise flat sheet, made for primarily decorative purposes.

Flame Types The various attributes of a torch flame as controlled by the relative amounts of air and fuel. A reducing flame has an excess of fuel, while an oxidizing flame has a surplus of air or oxygen. A flame in which all the fuel is being consumed is called neutral.

Fold forming A process of creating form and surface effects through a sequence of fold-flatten-anneal-open.

Generator curve A hypothetical line at right angles to the axial line, which defines the central axis of a stake.

Granulation An ancient technique in which tiny spheres of metal are fused to a parent sheet with fillets so tiny they are not visible without magnification.

Hardening & tempering The process of creating a different and dramatically tougher crystal in certain steels through controlled application of heat and cooling rates.

Kum-Boo The Korean name for a process of diffusing thin sheets of pure gold onto silver alloys.

Mallet A hammer-like tool made of wood, plastic, horn or a similar material that is used to manipulate metal without marring or thinning it.

Monoshell A hollow, volumetric form made from a single sheet of metal.

Mordant (etchant) A caustic oxidizing agent used in metalworking to selectively remove or texture sheets for decorative purposes. Nitric acid or ferric chloride are the two mordants most commonly used in jewelrymaking.

Neutralize The addition of a base to an acid or an acid to a base to bring the pH to a midrange balance. A typical application of the term refers to the neutralization of an acid pickle with a base such as baking soda.

Niello A blue-black alloy made of silver, copper and lead, mixed with sulfur to create a material that is laid into recesses for decorative effect.

Oxidized The result of oxygen combining with a material. In the case of metalworking, several common oxide films are called by familiar names; rust on steel, tarnish on sterling and patina on copper.

Patination The process or the result of a chemical reaction on the surface of a metal. Though the process happens whenever certain metals are exposed to air, the most common usage applies to controlled application of solutions intended to have a specific effect.

Pickle An acidic solution used in metalworking to chemically remove surface films such as oxides from a metal object.

Plain carbon steel A binary alloy in which the only addition to iron is carbon. This is contrasted with many other alloys in which several additional metals are added to contribute special prcperties to the alloy.

Planishing hammer A hammer usually of medium weight with polished faces that are used to smooth out a metal form. The word is derived from the same base as "planar."

Punch 1. A small simple tool used to decorate metal by creating an indentation. An example is a centerpunch. 2. The positive or male unit of a die system that is used to press a metal sheet into a form.

Quenchant The liquid used to cool a tool in the process of hardening or tempering. A thin light petroleum oil is the most common quenchant.

Resist An acid-proof paint that is used to protect selected areas from attack by the mordant (acid) when etching.

Reticulation The somewhat controlled process of wrinkling a surface by causing irregular expansion and contraction rates. The term also refers to the results of this process.

Rivet A rod that binds two or more units together by pressing them between flared out heads on each end.

Scoring The creation of grooves in a metal sheet, either through stock removal or compression, that will allow the work to be bent with a clean even crease.

Scotch-Brite® A plastic web that is used to clean metal. It is available in several coarsenesses and can be obtained with or without an abrasive charge.

Scratchbrush A hand or motor-driven brush of fine brass or stainless steel wires that is used to clean and polish metal. A light soap solution is always used when scratch-brushing to facilitate the burnishing action of the tool.

Silhouette die A one-piece die that determines the outline but not the contours of a form.

Sinusoidal stake A tapered steel rod that is shaped into a sequence of S-curves and is used in anticlastic raising.

Spiculum A tapering monoshell.

Synclastic A form in which the axial and generator lines both curve in the same direction. A bowl, for instance, is a synclastic form.

COLOPHON

Metals Technic was designed by Richard Mehl of Melilli.Mehl Graphic Designers in Portland, Maine in 1992. The pages were composed on a Macintosh using Quark 3.1. The illustrations were created by Tim McCreight. The photographs which open each chapter were made by Mary Melilli.

The typography used for the text is 10 point Memphis Light. Chapter titles are set in 22 point Memphis Bold, headings in 10 point Copperplate 33 Bold Condensed, and captions in 8 point Memphis Light.